T0257772

Ecological Water Treatment and Management

Ecological Water Treatment and Management

Edited by **Herbert Lotus**

New York

Published by Callisto Reference,
106 Park Avenue, Suite 200,
New York, NY 10016, USA
www.callistoreference.com

Ecological Water Treatment and Management
Edited by Herbert Lotus

International Standard Book Number: 978-1-63239-158-2 (Hardback)

Printed in the United States of America.

Contents

Preface

This book has been a concerted effort by a group of academicians, researchers and scientists, who have contributed their research works for the realization of the book. This book has materialized in the wake of emerging advancements and innovations in this field. Therefore, the need of the hour was to compile all the required researches and disseminate the knowledge to a broad spectrum of people comprising of students, researchers and specialists of the field.

The pollution of aquatic environment also reduces possible uses of water, especially those that require high quality standards i.e. for drinking purposes. This book discusses different problems regarding the quality of water and also focuses on the current techniques of water treatment. The primary emphasis is on sustainable choices of water usage that prevent water quality difficulties and aim at the protection of available water resources, and the development of the aquatic ecosystems. The focus is on the biological aspects of water quality using bioindices and biosensors.

At the end of the preface, I would like to thank the authors for their brilliant chapters and the publisher for guiding us all-through the making of the book till its final stage. Also, I would like to thank my family for providing the support and encouragement throughout my academic career and research projects.

Editor

Water Quality and Aquatic Ecosystems

Evaluation of Ecological Quality Status with the Trophic Index (TRIX) Values in the Coastal Waters of the Gulfs of Erdek and Bandırma in the Marmara Sea

Neslihan Balkis[1], Benin Toklu-Aliçli[1] and Muharrem Balci[2]
[1]Istanbul University, Faculty of Science, Department of Biology,
[2]Istanbul University, Institute of Science,
Vezneciler-Istanbul,
Turkey

1. Introduction

In developing countries, more than 90 percent of wastewater and 70 percent of industrial wastes are discharged into coastal waters without being treated (Creel, 2003). The entry of wastes into marine environment not only changes water quality parameters but also affects benthic organisms, causes habitat change and increases the risk of eutrophication and, thereby, causes the area to become susceptible. The Urban Wastewater Treatment Directive (UWTD; EC, 1991) defines eutrophication as the "enrichment of water by nutrients, especially compounds of nitrogen and/or phosphorus, causing an accelerated growth of algae and higher forms of plant life to produce an undesirable disturbance to the balance of organisms present in the water and to the quality of the water concerned". Karydis (2009) characterized "oligotrophic" waters as nutrient poor with low productivity, "eutrophic" waters as nutrient rich with high algal biomass and "mesotrophic" waters as moderate conditions. Hypoxia or even anoxia is the last stage of eutrophication (Gray, 1992) and this phase is often characterized as "dystrophic" (Karydis, 2009). In addition, eutrophication of coastal waters has been considered as one of the major threats to the health of marine ecosystems in the last few decades (Andersen et al., 2004; Yang et al., 2008). The risk of eutrophication may increase or decrease depending on the speed and direction of flow and wind. It can occur as a result of natural processes, for example, where there is upwelling of nutrient rich deep water to nutrient poor but light rich surface water of the photic zone of the water column (Jørgensen & Richardson, 1996). Cultural eutrophication arising from anthropogenic activities is particularly evident in marine areas with limited water exchange, and in lagoons, bays and harbours (Crouzet et al., 1999).

Various factors may increase the supply of organic matter to coastal systems, but the most common is clearly nutrient enrichment. The major causes of nutrient enrichment in coastal areas are associated directly or indirectly with meeting the requirements and demands of human nutrition and diet. The deposition of reactive nitrogen emitted to the atmosphere as a consequence of fossil fuel combustion is also an important anthropogenic factor (Nixon, 1995). Nutrients are the essential chemical components of life in marine environment. Phosphorus and nitrogen are incorporated into living tissues, and silicate is necessary for the formation of the skeletons of diatoms and radiolaria (Baştürk et al., 1986). In the sea, most of the nutrients are present in sufficient concentration, and lack of some of them limits the growth of phytoplankton (Pojed & Kveder, 1977). While some nutrient enrichment may be beneficial, excessive enrichment may result in large algal blooms and seaweed growths, oxygen depletion and the production of hydrogen sulphide, which is toxic to marine life and can cause high mortality, red tide events, decreasing fishery yields, and nonreversible changes in ecosystem health (Daoji & Daler, 2004).

Trophic conditions of European coastal waters vary considerably from region to region and within regions. A trophic index (TRIX) characterizing eutrophic levels, was introduced by Vollenweider et al. (1998). The European Environmental Agency has evaluated this index and suggested that TRIX scales at regional level should be developed. TRIX values are very sensitive and any slight change of oxygen, chlorophyll a, dissolved inorganic nitrogen and total phosphorus concentrations results in changed index values (Boikova et al., 2008). This simple index seems to help synthesize key eutrophication variables into a simple numeric expression to make information comparable over a wide range of trophic situations (Anonymous, 2001).

Bays and gulfs are very important for fishing since they provide habitats for sheltering and reproduction for most living species. They are influenced by environmental conditions, especially pollution, very rapidly, which causes negative changes in their structures. Bays and gulfs are quieter compared to open seas and have a semi-closed structure, which increases the frequency of such events as eutrophication and red-tide events. The influence level of pollution on living organisms is directly related with the changes in species diversity, and the effects of pollution on a specific environment can be determined by monitoring the natural process. However, in order to achieve this, the most important requisite is to determine the ecological status of the area(s) that will be studied before pollution (Koray & Kesici, 1994). Seasonal changes and global warming considerably affects the biological structure of seas (Goffart et al., 2002). These effects in the marine environment come into being with different phenomena. For instance, mucilage formation in seas is the aggregation of organic substances that are produced by various marine organisms under special seasonal and trophic conditions (Innamorati et al., 2001; Mecozzi et al., 2001). In Turkish territorial waters, mucilage formation was observed firstly in the Gulf of İzmit in the Marmara Sea in October 2007 and mainly fisheries and tourism have been damaged seriously (Tüfekçi et al., 2010). Then, mucilage formations were reported on the shores of Büyükada Island (Balkıs et al., 2011) and in the Gulf of Erdek (Tüfekçi et al., 2010). This study is important because there is not a sufficient amount of comprehensive research conducted on the subject in the Gulfs of Erdek and Bandırma. Besides, the two gulfs are important for fishing and they are under the threat of heavy pollution.

In recent years, the scientific and technological advances have shown that studying sea and oceans, which cover 70% of the earth, is considerably important. Today in order to meet the increasing need for food, the studies on food sources in our seas have gained speed. This study aims to determine the ecological quality of coastal waters in the Gulfs of Erdek and Bandırma, and firstly the two gulfs will be compared in terms of environmental factors.

Evaluation of Ecological Quality Status with the Trophic Index (TRIX) Values in the Coastal Waters of the Gulfs
of Erdek and Bandırma in the Marmara Sea

5

2. Material and methods

2.1 Study areas

The Marmara Sea forms "the Turkish Straits System" together with the Bosphorus and the Dardanelles. It has a surface area of 11,500 km² and the maximum depth is 1390 m. It is a small basin located between Asia and Europe. It is connected on the northeast with the Black Sea through the Bosphorus and on the southwest with the Aegean Sea through the Dardanelles (Ünlüata et al., 1990; Yüce & Türker, 1991; Beşiktepe et al., 1994). The brackish Black Sea waters with a salinity of about 17.6 ppt flow through the Bosphorus towards the Marmara Sea at the surface while the waters of Mediterranean origin with a salinity of about 38 ppt flow through the Dardanelles towards the Marmara Sea in a lower layer. There is an intermediate salinity mass with 25 m depth between these two water masses due to the difference in their densities (Ullyott & Pektaş, 1952; Yüce & Türker, 1991). The bottom water with a high salinity value includes a low amount of dissolved oxygen, and the water exchange and oceanographic conditions in the Marmara Sea are controlled by the two straits. The density stratification in the halocline impedes oxygen transfer from the surface layer that is rich in oxygen to the lower layer. Besides, biogenic particles in the bottom water increase oxygen consumption, which decreases the dissolved oxygen content of the lower water layer (Yüce & Türker, 1991; Beşiktepe et al., 2000).

The annual volume influx from the Black Sea to the Marmara Sea is nearly twice the salty water outflux from the Marmara basin to the Black Sea via the Bosphorus undercurrent (Ünlüata et al., 1990; Beşiktepe et al., 1994). Cyclonic alongshore currents in the Black Sea carry the polluted surface waters of the northwestern shelf as far as the Bosphorus region, with modified hydrochemical properties (Polat & Tuğrul, 1995). In addition, the salinity and nutrient contents of inflow slightly increase at the southern exit of the Bosphorus due to vertical mixing of the counter flows during the year. Concomitant photosynthetic processes in the Marmara upper layer, however, lead to consumption of biologically available nutrients and thus to a net export of particulate nutrients to the lower layer (Tuğrul et al., 2002). Primary production in the Sea is limited with halocline layer including the interface between 15 and 25 m (Polat & Tuğrul, 1995).

The oceanographic characteristics of the Gulfs of Erdek and Bandırma are similar to the Marmara Sea and the water column has a two-layer structure. The Gulf of Erdek has lower population density and industrial activities compared to the Gulf of Bandırma. The Biga River and the Gönen River flow into the Gulf of Erdek. The load of both rivers are affected by the mineral deposits and agricultural and food industries in their basin and the domestic wastes from the Boroughs of Biga and Gönen.

The Gulf of Bandırma is affected by industrial pollution and is more densely inhabited. The studies showed that the surface waters of the Gulf of Bandırma and the region that is on the northeast of the Kapıdağ Peninsula include more phosphate compared to the other parts of the Marmara Sea. This increase is caused by the domestic wastes and especially the fertilizer plant located in the Gulf. The Borough of Bandırma is rich in regards to nutrients both surface and ground waters. Most of the surface waters in Bandırma flow into the Susurluk River through Lake Manyas and the Kara River and reach the Marmara Sea. The most important harbor on the south of the Marmara Sea is located in this gulf. Although the intensive production of white meat and fertilizer raises the importance of the borough, it at the same time affects the Gulf of Bandırma negatively (Özelli & Özbaysal, 2001; Koç, 2002).

2.2 Sampling and primary analysis

Samples were collected from different depths of the water column (0.5, 15, 30 m) at three stations in each gulf, in total at six stations, seasonally (November, February, May, and August) for two years between November 2006 and August 2008 (Fig.1). The maximum depth at the stations is approximately 50 m. Photic zone, where photosynthesis occurs, was used as base in the selection of sampling depth. Water transparency was usually measured with the Secchi disk. A 3 l Ruttner water sampler with a thermometer was used for water analyses at each sampling point. The salinity was determined by the Mohr-Knudsen method (Ivanoff, 1972) and the dissolved oxygen (DO) by the Winkler method (Winkler, 1888).

Water samples for the determination of nutrients were collected in 100-mL polyethylene bottles and stored at -20 °C until the analysis in the laboratory. Nitrate+nitrite-N concentrations were measured by the cadmium reduction method using a "Skalar" autoanalyzer (APHA, 1999). Phosphate-P, Silicate-Si and chlorophyll a were analyzed by the methods described by Parsons et al. (1984). Chlorophyll a was measured after filtering 1 liter of the sample through Whatman GF/F filters. One milliliter of a 1% suspension of $MgCO_3$ was added to the sample prior to filtration. Samples were stored in a freezer, and pigments were extracted in a 90% acetone solution and measured with a spectrophotometer.

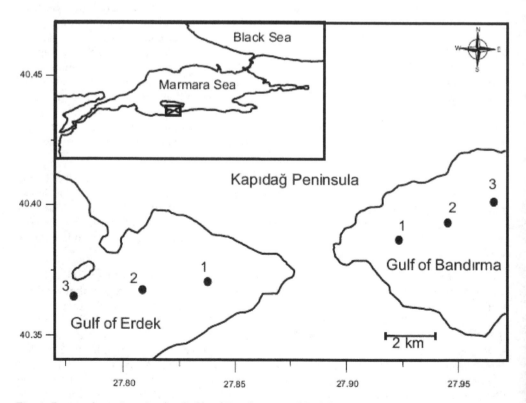

Fig. 1. Research stations in the Gulfs of Bandırma and Erdek

Evaluation of Ecological Quality Status with the Trophic Index (TRIX) Values in the Coastal Waters of the Gulfs
of Erdek and Bandırma in the Marmara Sea

7

2.3 Data analysis

Trophic Index (TRIX) values were calculated in order to determine the eutrophication level of the sampling area and the quality of waters (Vollenweider et al., 1998). The index is given by:

$$TRIX= [\log_{10}(Chl\ a \cdot D\%O \cdot N \cdot P)+1.5]/1.2$$

Chl a= Chlorophyll a (μg L^{-1}), D%O= Oxygen as an absolute deviation (%) from saturation, N= Dissolved inorganic nitrogen N-NO$_3$+NO$_2$ (μg-at L^{-1}), P= Total phosphorus P-PO$_4$ (μg-at L^{-1}). Ammonium values were not used in nutrient ratios and calculation of TRIX, because NH$_4$-N values were not measured in this paper. TRIX was scaled from 0 to 10, covering a range of four trophic states (0-4 high quality and low trophic level; 4-5 good quality and moderate trophic level; 5-6 moderate quality and high trophic level and 6-10 degraded and very high trophic level) (Giovanardi & Vollenweider., 2004; Penna et al., 2004).

Spearman's rank correlation coefficient (Siegel, 1956) was used to detect any correlation among biotic (chlorophyll a) and abiotic variables (temperature, salinity, DO and nutrients), and Bray-Curtis similarity index in Primer v6 software, based on log(x+1) transformation was calculated to detect the similarity between sampling stations (Clark & Warwick, 2001).

3. Results

The vertical distribution of temperature, salinity and dissolved oxygen in the coastal waters of the Gulfs of Bandırma and Erdek is shown in Figs. 2, 3. During the study, temperature, salinity and dissolved oxygen levels of the seawater ranged between 6.5 and 26 °C, 21.4 and 38.6 ppt, and 3.5 and 15.62 mg L^{-1} in the gulfs, respectively. Also, chlorophyll a values ranged between 0.1 and 14.79 μg L^{-1} (Figs. 2, 3).

In the Gulf of Bandırma, the highest temperature value (26 °C) was measured at the depth of 0.5 m at all stations (Fig. 2). Homogenous distribution of water temperature was observed at Station 1 in February 2007. Also, sudden changes were more pronounced after the depth of 15 m at all stations. In this Gulf, the highest salinity value (38.5 ppt) was determined at the depth of 30 m at Station 3 in November 2006. While upper layer salinity values were low, sudden increases were observed after the depth of 15 m. Dissolved oxygen content of the gulf was lower in the deeper layer compared to the upper. The highest DO value (15.62 mg L^{-1}) was measured at the depth of 15 m at Station 2 in November 2006, and the lowest (3.5 mg L^{-1}) at the depth of 30 m at Station 1 in May 2008. For chlorophyll a concentration, the highest value (14.79 μg L^{-1}) was determined at the depth of 0.5 m at Station 2 in August 2008 and higher chlorophyll a value was observed in the upper water column. The lowest value was detected at the depth of 30 m (0.21 μg L^{-1}) at Station 2 in May 2008. Water transparency was 4.5 m (February 2008) - 16 m (November 2006) in the Gulf of Bandırma.

In the Gulf of Erdek, the highest temperature value (25.5 °C) was recorded in the surface water at all stations in August 2007. Sudden temperature changes were detected after the depth of 15 m at all stations as in the Gulf of Bandırma. Salinity values ranged from 22.4 ppt (st.2, 0.5 m, May 2007) to 38.6 ppt (st.3, 30 m, November 2006). While a sudden increase was detected in salinity values after the depth of 15 m at Station 2 and Station 3, the values increased in some seasons and decrease in others at Station 1, which is a coastal station. As in the Gulf of Bandırma, oxygen values were determined to be higher in the upper layer and showed decrease towards the deeper layers. The lowest value (3.78 mg L⁻¹) was detected at the depth of 30 m at Station 3 in August 2008. During most of the sampling period, chlorophyll a values were higher in the euphotic zone and showed decrease in correlation with the increase of depth. Chlorophyll a values were between 0.10 μg L⁻¹(st.3, 30 m, May 2008) and 2.83 μg L⁻¹ (st.3, 15 m, February 2008). Water transparency was 6 m (February 2008) - 15 m (February 2007) in this gulf.

Nutrient concentrations and TRIX index values are shown in Figs. 4, 5. The amounts of nitrate+nitrite-N (0.07-5.83 μg-at N L⁻¹), phosphate-P (0.09-8.6 μg-at P L⁻¹) and silicate-Si (0.05-21.62 μg-at Si L⁻¹) concentrations were measured. The consumption of nitrogen and silica in the upper layer was determined to be higher in both of the gulfs. However, phosphorus values were quite high in the upper layer compared to the lower especially in August 2008. A similar situation was observed at Station 1 in February 2008. There were low levels of dissolved oxygen in the deeper layers, which were rich in nutrients.

Mean ratios of nutrients and chlorophyll a at the sampling stations are presented in Table 1. The lowest and highest mean ratios of N/P were 0.04 (0.5 m depth, in May 2007) and 4.73 (30 m depth, in May 2007) in the Gulf of Bandırma and 0.35 (0.5 m depth, in August 2007) and 7.08 (30 m depth, in May 2008) in the Gulf of Erdek, respectively. Also these ratios were recorded lower than the Redfield ratio (16/1). This result indicates that N is limiting nutrient for both gulfs. A considerable increase was observed in both N/P and Si/P ratios especially from the upper layer to the lower in both gulfs. Besides, these values were higher in the Gulf of Erdek compared to the Gulf of Bandırma. In both gulfs, N/Si ratio was lower than 1 during all sampling periods. P/Chl a ratio was low in the upper layer due to the increase of chlorophyll a value depending on phytoplankton activity and the use of phosphorus by these organisms.

Period	N/P			N/Si			Si/P			P/Chl a		
	0.5m	15m	30m	0.5m	15m	30m	0.5m	15m	30m	0.5m	15m	30m
Gulf of Bandırma												
Nov.2006	0.34	0.35	3.42	0.21	0.26	0.29	1.64	1.49	11.22	1.12	0.87	2.88
Feb.2007	0.35	0.13	3.82	0.19	0.07	0.52	1.80	1.81	7.76	0.66	0.70	3.28
May.2007	0.04	0.33	4.73	0.17	0.26	0.34	0.28	1.30	13.30	1.18	1.38	2.99
Aug.2007	0.65	0.23	3.18	0.45	0.21	0.27	4.20	1.81	11.39	0.98	0.80	1.89
Nov.2007	0.68	1.26	2.40	0.30	0.29	0.55	1.98	3.71	4.52	0.37	0.55	1.78
Feb.2008	0.57	0.47	3.08	0.19	0.24	0.70	3.29	1.90	4.39	0.57	0.19	0.49
May.2008	0.48	3.28	4.37	0.77	0.80	0.75	0.60	3.66	5.77	3.00	1.68	3.84
Aug.2008	0.08	0.08	2.37	6.25	2.38	1.35	0.02	0.17	3.19	0.64	0.58	0.72
Gulf of Erdek												
Nov.2006	0.68	2.32	4.70	0.12	0.30	0.36	6.05	7.33	12.88	0.25	1.18	1.91
Feb.2007	0.91	1.00	3.24	0.17	0.10	0.14	5.54	16.05	23.62	0.31	0.68	4.39
May.2007	0.85	0.55	3.70	0.13	0.07	0.21	6.45	13.11	17.78	0.19	0.25	2.06
Aug.2007	0.35	0.77	1.82	0.06	0.14	0.10	5.69	5.30	13.67	0.61	0.52	0.37
Nov.2007	0.72	1.47	0.97	0.44	0.66	0.49	1.42	1.98	2.62	0.61	1.02	6.42
Feb.2008	1.91	1.70	3.22	0.15	0.13	0.43	12.88	13.51	8.97	0.06	0.05	0.13
May.2008	1.80	2.33	7.08	0.92	0.54	0.75	1.93	4.60	9.42	0.35	0.52	4.93
Aug.2008	1.12	0.71	5.09	0.55	0.44	0.54	2.07	1.44	9.30	0.19	0.55	0.75

Table 1. The atomic ratios of nutrients and chlorophyll a at the sampling stations.

Evaluation of Ecological Quality Status with the Trophic Index (TRIX) Values in the Coastal Waters of the Gulfs of Erdek and Bandırma in the Marmara Sea

9

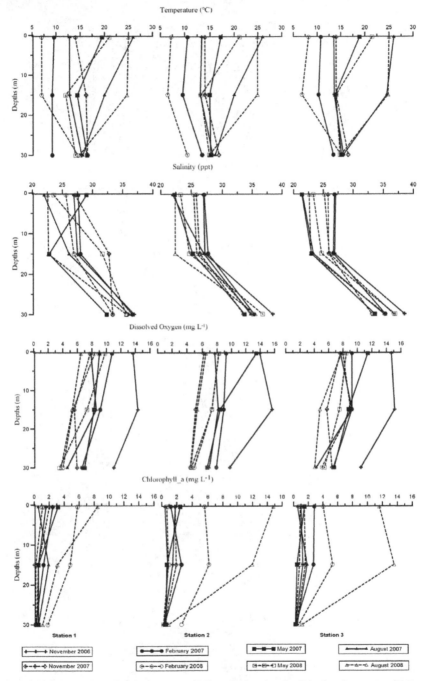

Fig. 2. Vertical variations of temperature (°C), salinity (ppt), dissolved oxygen (DO, mg L⁻¹) and chlorophyll *a* (μg L⁻¹) along the water column in the Gulf of Bandırma.

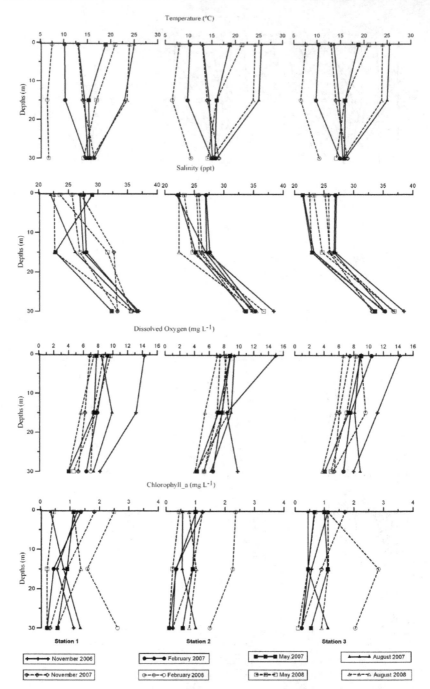

Fig. 3. Vertical variations of temperature (°C), salinity (ppt), dissolved oxygen (DO, mg L^{-1}) and chlorophyll a (µg L^{-1}) along the water column in the Gulf of Erdek.

Evaluation of Ecological Quality Status with the Trophic Index (TRIX) Values in the Coastal Waters of the Gulfs of Erdek and Bandırma in the Marmara Sea

11

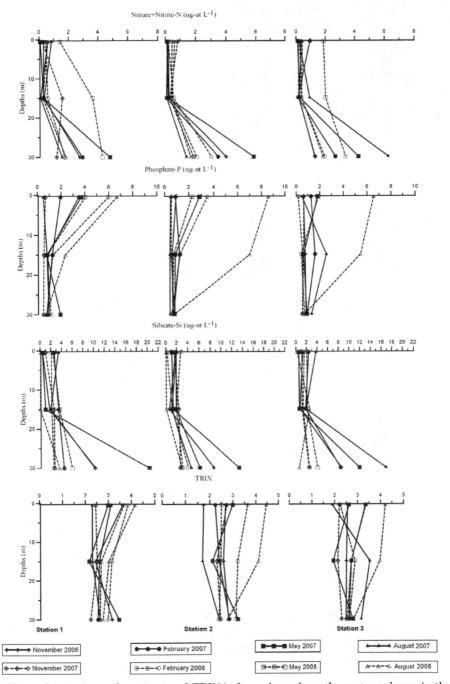

Fig. 4. Vertical variations of nutrient and TRIX index values along the water column in the Gulf of Bandırma.

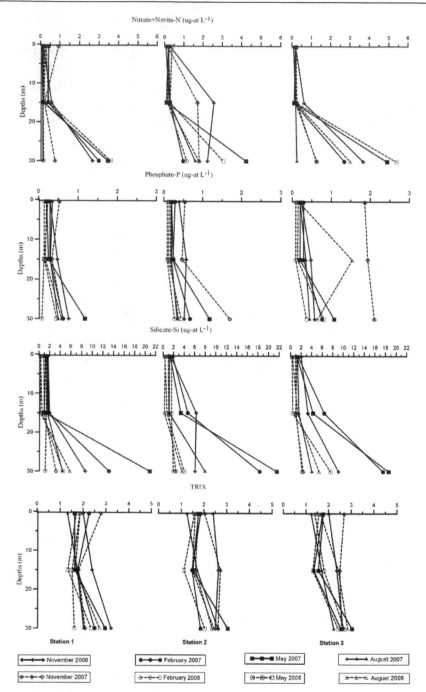

Fig. 5. Vertical variations of nutrient and TRIX index values along the water column in the Gulf of Erdek.

Evaluation of Ecological Quality Status with the Trophic Index (TRIX) Values in the Coastal Waters of the Gulfs of Erdek and Bandırma in the Marmara Sea

13

While the Trophic Index (TRIX) value for the Gulf of Erdek was determined to range between 1.12 and 3.23, it ranged between 1.68 and 4.46 for the Gulf of Bandırma (Figs. 4, 5). In addition, an increase was observed in TRIX values with the increase in concentrations of phosphorus and chlorophyll a in the last season (August 2008) of the second sampling period in the Gulf of Bandırma.

The results of Spearman's rank order correlation were employed to explain the relationship among the ecological parameters in the gulfs (Table 2, 3). The nutrients were negatively correlated to dissolved oxygen, but positively to salinity except phosphorus in the Gulf of Bandırma. Also, chlorophyll a was negatively correlated with N and Si in the gulfs, however it was positively correlated with P in the Gulf of Bandırma and negatively in the Gulf of Erdek.

	Dissolved oxygen	Temperature	Salinity	Chlorophyll a	Nitrogen	Phosphorus	Silica
Temperature	-,190						
Salinity	-.274**	-.279**					
Chlorophyll a	,139	-,132	-.545**				
Nitrogen	-.482**	,020	.604**	-.521**			
Phosphorus	,170	,106	-.238*	.443**	-,055		
Silica	-.247*	-,175	.706**	-.579**	.699**	-,106	
TRIX	,022	-,043	-,022	.331**	.355**	.640**	,118

Table 2. Spearman's rank-correlation matrix (r_s) to correlate among ecological variables in the Gulf of Bandırma (** $P<0.01$, * $P<0.05$, n=72)

	Dissolved oxygen	Temperature	Salinity	Chlorophyll a	Nitrogen	Phosphorus	Silica
Temperature	-,071						
Salinity	-.390**	-.360**					
Chlorophyll a	,151	-.335**	-.280**				
Nitrogen	-.455**	-,117	.675**	-.448**			
Phosphorus	-.397**	,005	.625**	-.345**	.593**		
Silica	-.266*	-.216*	.751**	-.336**	.525**	.618**	
TRIX	-.365**	-.206*	.653**	,000	.745**	.807**	.551**

Table 3. Spearman's rank-correlation matrix (r_s) to correlate among ecological variables in the Gulf of Erdek (** $P<0.01$, * $P<0.05$, n=72)

The Bray-Curtis similarity index did not show significant differences at the sampling stations according to ecological parameters. Sampling stations in the gulfs were approximately 91% similar to each other (Fig. 6).

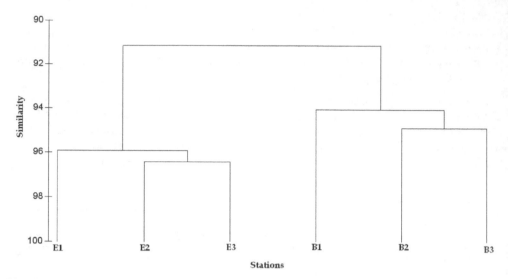

Fig. 6. Bray-Curtis similarity dendogram of the sampling stations in the Gulfs of Bandırma and Erdek.

4. Discussion

The chemical oceanography of the Marmara Sea is remarkably affected by the Black Sea and the Aegean Sea, and the basin includes two different masses. In this study, the highest temperature values were measured at the depth of 0.5 m in August 2007 in both gulfs with similar characteristics. Especially, the variations of temperatures in surface water showed typically seasonal trend and this was caused by the effect of light and the contact of this layer with the atmosphere. While lower values were measured in the lower layer water compared to the upper in spring and summer, temperature values increased in correlation with the increase in depth in autumn and winter. The increase in temperature at the depth of 30 m during cold periods indicates the effect of the Mediterranean waters. Especially, a sharp decrease was detected in water temperature after the depth of 15 m in August in 2007 and 2008 during all sampling periods in the both gulfs. It was found that less saline water from the Black Sea via the Bosphorus was effective at the depths that were close to the surface and salinity was noted to increase from the surface to the bottom, reaching its highest value at the depth of 30 m due to the Mediterranean current. After 15 m, a sudden increase in salinity was remarkable, which indicates the presence of a halocline layer. Besides, in gulfs salinity values changed seasonally throughout the water column at Station 1, which is an inner one while seasonal changes were more stable until the depth of 15 m at Station 2 and 3 and the values showed sudden increases after this depth.

Rather high oxygen concentrations were observed in the upper water, probably coinciding with the maximum of the photosynthetic activity. A decrease was observed in dissolved oxygen values generally from the surface to the bottom along the water column. This was due to excessive oxygen consumption during the decomposition of detritus, which was produced as a consequence of primary production in the upper layer and biochemical reactions occurring in the deeper layer. Excessive bacterial and animal activity due to increased phytoplankton biomass and high organic loads in eutrophic systems can lead to oxygen depletion (Karydis, 2009). This decrease in the water column was in accordance with the result of the previous review study (Yılmaz, 2002). The most remarkable period in terms of seasonal changes in oxygen was November 2006 in both gulfs. Considering the other sampling periods, the highest dissolved oxygen values were recorded at all depths in this sampling period. The fate and behavior of DO is of critical importance to marine organisms in determining the severity of adverse impacts (Best et al., 2007). When the DO falls below 5 mg L^{-1}, sensitive species of fish and invertebrates can be negatively impacted, and at the DO levels below 2.5 mg L^{-1} most fish are negatively impacted (Frodge et al., 1990). Best et al. (2007) provided DO thresholds in accordance with 5 ecological categories in the European Water Framework Directive ($\geq$5.7 mg L^{-1} High; $\geq$4.0<5.7 mg L^{-1} Good; $\geq$2.4<4.0 mg L^{-1} Moderate; $\geq$1.6<2.4 mg L^{-1} Poor; <1.6 mg L^{-1} Bad). In our study, high and good quality status were detected in both gulfs and moderate quality status were only detected at the depth of 30 m at Station 1 in May and August 2008 in the Gulf of Bandırma and at Station 3 in August 2008 in the Gulf of Erdek. Especially, salinity at 30 m, which increase with the effect of the Mediterranean waters, showed a negative correlation with DO and water quality is moderate at this depth according the DO levels. The main physical factors affecting the concentration of oxygen in the marine environment are temperature and salinity: DO solubility decreasing with increasing temperature and salinity (Best et al., 2007). Negative correlation was found between DO and salinity in the present study (p<0.01). Although pollution has clearly increased in last 30 years in the upper water of the Marmara Sea, dissolved oxygen values in the deeper layer have not changed compared to the values measured in 1970s (Tuğrul et al., 2000).

Chlarophyll *a* production and nutrient availability are closely associated with eutrophication (Nixon, 1995; Kitsiou & Karydis, 2001). Chlorophyll *a* distribution depends on hydro-chemical conditions, namely nutrient availability, temperature changes, light conditions, water turbulence etc. (Lakkis et al., 2003; Nikolaidis et al., 2006 a, b). In this study, the highest chlorophyll *a* values were generally determined in winter period in both gulfs and P/Chl *a* ratio was low in the upper layer due to the increase of chlorophyll *a* value depending on phytoplankton activity and the use of phosphorus by these organisms; however, chlorophyll *a* showed excessive increase only in summer (August 2008) in the Gulf of Bandırma. Especially a serious environmental problem was observed in 2008 in the whole Marmara Sea. In recent studies, it was stated that mucilage formation, which was observed mainly due to excretory activity of some diatoms together with bacteria, the dinoflagellate *Gonyaulax fragilis*, the presence of sharp picnocline and thermocline caused by the two-layered water system of the Marmara Sea in 2008; besides, which the weather conditions and the status of currents during that time effected this formation (Tüfekçi et al., 2010; Balkıs et al., 2011). During the mucilage formation chlorophyll *a* values changed between 0.1-22 µg L^{-1} in these studies. According to the study by Ignatiades (2005), the limits of average concentration in chlorophyll *a* are <0.5 µg L^{-1} for oligotrophic, 0.5-1.0 µg L^{-1} for mesotrophic and >1.0 µg L^{-1} for eutrophic waters. According to chlorophyll *a* results obtained from both gulfs, the gulfs showed mesotrophic conditions in some periods and eutrophic, hyper-eutrophic (during the mucilage formation event) conditions in others.

Fiocca et al. (1996) reported that the availability of dissolved inorganic nitrogen and dissolved inorganic phosphorus leads to a seasonal change in N/P ratio, high in winter and low in summer. In the euphotic zone, nutrients, especially nitrate+nitrite-N and silicate-Si, are practically depleted by the phytoplankton uptake. In the Marmara Sea, the highest abundance of phytoplankton was recorded in the surface water (0.5-5 m) in recent comprehensive studies (Balkıs, 2003; Deniz & Taş, 2009; Tüfekçi et al., 2010; Taş et al., 2011). The highest nutrient values were recorded generally in the bottom layer where there was aggregation. During the study, positive correlations (p<0.01) of nitrogen and silicate were detected with salinity, which increased with depth. The strong positive correlation (p<0.01) between nitrogen, phosphate and silicate in the Gulf of Erdek and between nitrogen and silicate in the Gulf of Bandırma might indicate that these nutrients come from the same sources into the water column. Especially, the increase in the amount of nitrogen and silica at the depth of 30 m is remarkable since these elements are produced by the bacterial decomposition of the organic substances aggregating at the bottom.

Smith (1984) mentioned that nitrogen budget in the ocean was associated with air-water interaction such as N_2 fixation, losses of fixed nitrogen, or sediment back to gaseous form but that it depends on availability of phosphorus because phosphorus is not exchanged between the ocean and atmosphere as nitrogen. In both gulfs, there was negative correlation (p<0.01) between chlorophyll a and nutrients except for the positive correlation (p<0.01) between chlorophyll a and phosphorus in the Gulf of Bandırma (Tables 2, 3). It is known that eutrophication variables such as nutrient and chlorophyll a do not seem to follow a linear relationship (Karydis, 2009). Interestingly, there was a negative correlation (p<0.05) between salinity and phosphate in the Gulf of Bandırma and positive correlation (p<0.01) in the Gulf of Erdek. The inverse relationship between phosphate and salinity occurs with increasing of evaporation, which depends on temperature while remain nitrogen is at low level concentrations in the gulf (Smith, 1984). Actually, the positive correlation (p<0.01) between phosphate and salinity, which increased with depth indicates the source of phosphate in the water column in the Gulf of Erdek. Smith (1984) argued that the sediments could be reliable records for net nitrogen and phosphorus accumulation in the bays, which have very slow water turnover. Generally, nutrients are depleted by phytoplankton at the points where the light reaches while it increases in direct proportion to depth. The negatively correlated relationships between chlorophyll a and nutrients might indicate that nutrients are controlled by primary producers in the short term (Pérez-Ruzafa et al., 2005). In terms of phosphorus inputs, the significant positive relationships between phosphorus and chlorophyll a might indicate that the Gulf of Bandırma has not reached the saturation level, which has an enhanced effect on primary productivity. Jaanus (2003) asserted that phosphorus was more important on primary production in eutrophied coastal areas rather than nitrogen when a positive correlation between phosphorus and chlorophyll a was detected. On the other hand, the significant negative relationship between nutrients and chlorophyll a probably indicates that these nutrients have excessive concentrations, which have a restricted effect on primary production with regard to phosphorus in the Gulf of Bandırma.

Many authors are of the opinion that it is useful to look at the N/Si/P ratios in various parts of the ocean, and that only certain ratios are favorable for bioproductivity. It is known that P stress is common in freshwater systems, whereas N stress is found in marine systems (Ryther & Dunstan, 1971). The nitrogen limitation of phytoplankton growth is common in coastal systems (Nixon, 1986). Redfield et al. (1963) mentioned a C/N/P ratio of 106/16/1 ratio among the elements of sea water. If N/P ratios are below the Redfield value of 16/1, N is limiting nutrient (Stefanson et al., 1963). In the coastal waters of the Gulfs of Bandırma and Erdek, the atomic ratio of N/P was lower than the Redfield ratio of 16, and N was limiting nutrient. The increase in the N/P ratio was remarkable in lower layers in comparison to the surface water, which showed that phosphorus increased in proportion to nitrogen in the surface water while nitrogen increased in proportion to phosphorus in lower layers. In addition, the reason for lower values of N/P ratio in summer months is the limiting effect of nitrogen (Marcovecchio et al., 2006) and in the eutrophic areas, nitrogen might be a significant growth-limiting factor under the conditions, which have high total phosphorus concentration and low total N/P ratio (Attayde & Bozelli, 1999). Diatom growth in marine waters is likely to be limited by dissolved silica when the N/Si ratios are above 1/1 (Roberts et al., 2003). During the study period, the N/Si ratios were low (<1), and this indicates that silicate is not a limiting factor, especially for the growth of diatoms. In the studies conducted in the Marmara Sea, it was reported that dinoflagellates and diatoms constitute most of the plankton population, which supports our finding (Balkıs, 2003; Deniz & Taş, 2009; Tüfekçi et al., 2010; Balkıs et al., 2011). It is known that especially diatoms show an excessive increase in summer and spring (Balkıs, 2003; Deniz & Taş, 2009).

The Secchi disk depth for oligotrophic waters varies from 20 to 40 m (Ignatiades et al., 1995). Secchi disk depth detected between 10 and 20 m characterizes mesotrophic conditions whereas it is less than 10 m for eutrophic waters (Ignatiades et al., 1995). The lowest-highest Secchi range recorded in this study is 4.5-16 m. The values measured in 2008 are lower than those recorded in the previous year. Especially in winter when environmental inputs are intense lower values were recorded. In addition, intense vertical mixtures also caused turbidity in this period.

Trophic condition of vast marine areas, like the Mediterranean, varies considerably from region to region and within regions (Vollenweider et al., 1996). Vollenweider et al. (1998) calculated TRIX mean values as 3.37-5.60 for the Adriatic Sea. Moncheva et al. (2001) detected that TRIX index values varied from 5.0 to 6.0 in the Thermaikos Gulf of the northern Aegean Sea, and the lowest values were recorded in summer. In addition, the index was recorded between 1.9-4.7 in Kalamitsi on the east coasts of the central Ionian Sea (Nikolaidis et al., 2008), 6.90-7.70 in southern Black Sea (Baytut et al., 2010), 0.86-2.98 in the Edremit Bay of the Aegean Sea (Balkıs & Balcı, 2010). Calculated TRIX values were slightly lower than expected because NH_4 was not measured in this study and used in the original formula. Low TRIX values, as defined in this paper, indicate poorly productive waters corresponding to high water quality in the Gulfs of Erdek and Bandırma. On the other hand, according to chlorophyll *a* results the environment generally showed mesotrophic-eutrophic conditions. The chlorophyll *a* scale used above (Ignatiades, 2005) is the one suggested for the Aegean Sea. The Marmara Sea, different from the Aegean Sea, is an inland sea and has a two-layered water system. The calculated Trix values without NH_4 could caused to obtained low values. Especially the upper layer waters are under the effect of the waters of Black Sea, which has intense river inputs. Therefore, in order to determine the water quality of the environment not only physical and chemical studies but also biological studies that will show the organism communities in the environment and their abundance should be conducted. This study showed the current state of water quality in both gulfs and can be used as a main source in determining possible changes in these gulfs in future. Moreover, effective wastewater management including nutrient control may be required for an effective pollution prevention program in these regions.

5. Acknowledgements

This study was supported by the Research Fund of Istanbul University, project number 541.

6. References

American Public Health Association (APHA) (1999). *Standard Methods for the Examination of Water and Waste Water*, 20th ed. Washington DC, USA.

Andersen, J.H., Conley, D.J. & Hedal, S. (2004). Palaeoecology, reference conditions and classification of ecological status: the EU Water Framework Directive in practice. *Marine Pollution Bulletin*, 49, 283-290.

Attayde, J. L. & Bozelli, R. L. (1999). Environmental heterogeneity patterns and predictive models of chlorophyll *a* in a Brazilian coastal lagoon. Hydrobiologia, 390 (1-3), 129-139.

Balkıs, N. (2003). Seasonal variations in the phytoplankton and nutrient dynamics in the neritic water of Büyükçekmece Bay, Sea of Marmara. *Journal of Plankton Research*, 25(7), 703-717.

Balkıs, N. & Balcı, M. (2010). Seasonal variations of nutrients and chlorophyll *a* in the coastal waters of the Edremit Bay and The trophic index (TRIX) values of the environment. *Fresenius Environmental Bulletin*, 19(7), 1328-1336.

Balkıs, N., Atabay, H., Türetgen, I., Albayrak, S., Balkıs, H. & Tüfekçi, V. (2011). Role of single-celled organisms in mucilage formation on the shores of Büyükada Island (the Marmara Sea). *Journal of the Marine Biological Association of the United Kingdom*, 91 (4), 771-781.

Baştürk, Ö., Saydam, A.C., Salihoğlu, İ. & Yılmaz, A. (1986). Oceanography of the Turkish Straits', *First Annual Report*, Vol. III: health of Turkish Straits, II. Chemical and environmental aspects of the Sea of Marmara. Institute of marine Sciences, METU, Erdemli-İçel, Turkey.

Baytut, O., Gonulol, A. & Koray, T. (2010). Temporal variations of phytoplankton in relation to eutrophication in Samsun Bay, Southern Black Sea. *Turkish Journal of Fisheries and Aquatic Sciences*, 10, 363-372.

Best, M.A., Wither, A.W. & Coates, S. (2007). Dissolved oxygen as a physico-chemical supporting element in the water framework directive. *Marine Pollution Bulletin.*, 55, 53-64.

Beşiktepe, Ş.T., Özsoy, E., Abdullatif, M.A. & Oğuz, T. (2000). Marmara Denizi'nin Hidrografisi ve Dolaşımı, *Marmara Denizi 2000 Sempozyumu Bildiriler Kitabı*, 11-12 Kasım 2000 Ataköy Marina, İstanbul, Yayın No: 5 Türk Deniz Araştırmaları Vakfı, İstanbul: 314-326.

Beşiktepe, Ş.T., Sur, H.İ., Özsoy, E., Abdullatif, M.A., Oğuz, T. & Ünlüata, Ü. (1994). The circulation and hydrography of the Marmara Sea, *Progress in Oceanography*, 34, 285-334.

Boikova, E., Botva, U. & Licite, V. (2008). Implementation of trophic status index in brackish water quality assessment of Baltic coastal waters. *Proceedings of the Latvian Academy of Sciences*, Section B, 62(3), 115-119.

Clarke, K.R. & Warwick, R.M. (2001). Change in marine communities: an approach to statistical analysis and interpretation, second ed., PRIMER-E: Plymouth.

Evaluation of Ecological Quality Status with the Trophic Index (TRIX) Values in the Coastal Waters of the Gulfs
of Erdek and Bandırma in the Marmara Sea

19

Creel, L. (2003). Ripple effects: Population and coastal regions. (policy brief). Washington, DC: Population Reference Bureau.

Crouzet, P., Nixon, S., Laffon, L., Bøgestrand, J., Lallana, C., Izzo, G., Bak, J. & Lack, T.J. (1999). Status and impact on marine and coastal waters, In: *Nutrients in European ecosystems,* Thyssen, N., (Ed.), pp. 81-93, Office for Official Publications of the European Communities, EEA, Copenhagen.

Daoji, L. & Daler, D. (2004). Ocean pollution from land-based sorces: East China Sea, China. *Ambio,* 33, 107-113.

Deniz, N & Taş, S. (2009). Seasonal variations in the phytoplankton community in the northeastern Sea of Marmara and a species list. *Journal of the Marine Biological Association of the United Kingdom,* 89(2), 269-276.

EC. 1991. Urban Wastewater Treatment Directive 91/271/EEC. Official Journal of the European Communities, L135/40–52, 30 May 1991.

Fiocca, F., Luglie, A. & Sechi, N. (1996). The phytoplankton of S'Ena Arrubia Lagoon (centrewestern Sardinia) between 1990 and 1995. *Giornale Botanico Italiano,* 130, 1016-1031.

Frodge, J.D., Thamas, G.L. & Pauley, G.B. (1990). Effects of canopy formation by floating and submergent aquatic macrophytes on the water quality of two shallow Pacific Northwest lakes. *Aquatic Botany,* 38, 231-248.

Giovanardi, F. & Vollenweider, R.A. (2004). Trophic conditions of marine coastal waters: experience in applying the Trophic index TRIX to two areas of the Adriatic and Tyrrhenian seas. *Journal of Limnology,* 63, 199-218.

Goffart, A., Hecq, J.H. & Legendre, L. (2002). Changes in the development of the winterspring phytoplankton bloom in the Bay of Calvi (NW Mediterranean) over the last two decades: a response to changing climate? *Marine Ecology Progress Series,* 236, 45-60.

Gray, S.J. (1992). Eutrophication in the sea. In: *Marine eutrophication and pollution dynamics,* Colombo, G. and Viviani, R. (Eds), Olsen&Olsen, Fredensborg, 394p.

Ignatiades, L. (2005). Scaling the trophic status of the Aegean Sea, eastern mediterranean. *Journal of Sea Research.,* 54, 51-57.

Ignatiades, L., Georgooulos D. & Karydis, M. (1995). Description of a phytoplanktonic community of the oligotrophic waters of SE Aegean Sea (Mediterranean), *P.S.Z.I. Marine Ecology,* 16, 13-26.

Innamorati, M., Nuccio, C., Massi, L., Mori, G. & Melley, A. (2001). Mucilages and climatic changes in the Tyrrhenian Sea. *Aquatic Conversation: Marine and Freshwater Ecosytems,* 11, 289-298.

Ivanoff, A. (1972). *Introduction al'océanographie,* Tome I. Librairie Vuibert, Paris.

Jaanus, A. (2003). Water environment of Haapsalu Bay in retrospect (1975-2000). Proceedings of the Estonian Academy of Sciences. Ecology, 52 (2), 91-111.

Jorgensen, B.B., Richardson, K. (Eds). (1996). *Eutrophication in coastal marine ecosystems.* Coastal and Estuarine Studies 52. American Geophysical Union, Washington, DC.

Karydis, M. (2009). Eutrophication assessment of coastal waters based on indicators: a literature review. *Global NEST Journal,* 11(4), 373-390.

Kitsiou, D. & Karydis, M. (2001). Marine eutrophication: a proposed data analysis procedure for assessing spatial trends, *Environmental Monitoring and Assessment*, 68, 297-312.

Koç, T. (2002). Bandırma ilçesinde tavukçuluğun çevresel etkisi, *Ekoloji*,11 (43), 11-16.

Koray, T. & Kesici, U. Y. (1994). Bodrum Körfezi'nin (Ege Denizi) fitoplankton ve protozooplankton tür kompozisyonu. *Ege Üniversitesi Fen Fakültesi Dergisi*, Seri B, Ek 16/1, 971-980.

Lakkis, S., Jonsson, L., Zodiatis, G. & Soloviev, D. (2003). Remote sensing data analysis in the Levantine basin: SST and chlorophyll *a* distribution. In: Oceanography of Eastern mediterranean and Black Sea. Similarities and differences of two interconnected basins (Ed. A. Yilmaz). 14-18 October, 2002, TUBITAK Publishers, Ankara, pp. 266-273.

Marcovecchio, J., Freije, H., De Marco, S., Gavio, A., Ferrer, L., Andrade, S., Beltrame, O. & Asteasuain, R. (2006). Seasonality of hydrographic variables in a coastal lagoon: Mar Chiquita, Argentina, Aquatic Conserv: Marine Freshwater Ecosystem, 16, 335-347.

Mecozzi, M., Acquistucci, R., Di Noto, V., Pietrantonio, E., Amirici, M. & Cardarilli, D. (2001). Characterization of mucilage aggregates in Adriatic and Tyrrhenian Sea: structure similarities between mucilage samples and the insoluble fractions of marine humic substance, *Chemosphere*, 44, 709-720.

Moncheva, S., Gotsis-Skretas, O., Pagou, K. & Krastev, A. (2001). Phytoplankton blooms in Black Sea and Mediterranean coastal ecosystems subjected to anthropogenic eutrophication: similarities and differences. *Estuarine Coastal Shelf Science*, 53, 281-295.

Nikolaidis, G., Moschandreou, K. & Patoucheas, D.P. (2008). Application of a trophic index (TRIX) for water quality assessment at Kalamitsi coasts (Ionian Sea) after the operation of the wastewater treatment plant. *Fresenius Environmental Bulletin*, 17(11), 1938-1944.

Nikolaidis, G., Patoucheas, D.P. & Moschandreou, K. (2006a). Estimating breakpoints of chl-a in relation with nutrients from Thermaikos Gulf (Greece) using piecewise linear regression. *Fresenius Environmental Bulletin*, 15 (9b), 1189 - 1192.

Nikolaidis, G., Moschandreou, K., Koukaras, K., Aligizaki, K., Kalopesa, E. & Heracleous, A. (2006b). The use of chlorophyll *a* for trophic state assessment of water masses in the inner part of Thermaikos Bay (Nw Aegean Sea). *Fresenius Environmental Bulletin*, 15 (9b), 1193 - 1198.

Nixon, S.W. (1986). Nutrient dynamics and the productivity of marine costal waters, In: *Coastal eutrophication* (Eds. R.D. Halwagy, B. Clayton, M. Behbehani), pp. 97-115, The Alden Press, Oxford.

Nixon, S.W. (1995). Coastal marine eutrophication: a definition, social causes, and future concerns. *Ophelia*, 41, 199-219.

Özelli C. & Özbaysal M.K. (2001). Kalkınmada öncelikli yöreler, Balıesir il profile, TC. Sanayi Bakanlığı KOSGEB.

Parsons, T.R., Maita, Y. & Lalli, C.M. (1984). A manual of chemical and biological methods for seawater analysis. Pergamon Press, U. K.

Penna, N., Capellacci, S. & Ricci, F. (2004). The influence of the Po River discharge on phytoplankton bloom dynamics along the coastline of Pesaro (Italy) in the Adriatic Sea. *Marine Pollution Bulletin*, 48, 321-326.

Perez-Ruzafa, A., Fernandez, A.I., Marcos, C., Gilabert, J., Quispe, J.I. & Garcia-Charton, J.A. (2005). Spatial and temporal variations of hydrological conditions, nutrients and chlorophyll *a* in a Mediterranean coastal lagoon (May Menor, Spain). Hydrobiologia, 550, 11-27.

Pojed, I. & Kveder, S. (1977). Investigation of nutrient limitation of phytoplankton production in the Northern Adriatic by enrichment experiments. *Thalassia Jugoslavica*, 13, 13-24.

Polat, Ç.S. & Tuğrul, S. (1995). Nutrient and organic carbon exchanges between the Black and Marmara seas through the Bosphorus strait, *Continental Shelf Research*, 15(9), 1115-1132.

Redfield, A.C., Ketchum, B.K. & Richards, F.A. (1963). The influence of organisms on the composition of sea water. In: *The sea*. Vol.2. (Ed. M.N. Hill), , pp. 26-77, Wiley, New York.

Roberts, E.C., Davidson, K. & Gilpin L.C. (2003). Response of temperate microplankton communities to N:Si ratio perturbation. Journal of Plankton Research, 25, 1485-1495.

Ryther, J.H. & Dunstan, W.M. (1971). Nitrogen, phosphorus and eutrophication in the coastal marine environment. Science, 171, 1008-1013.

Siegel, S. (1956). Non parametric statistics for the behavioral sciences. McGraw-Hill, New York.

Smith, S.V. (1984). Phosphorus versus nitrogen limitation in the marine environment. Limnology and Oceanography, 29, 1149-1160.

Stefanson, U. & Richards, F.A. (1963). Processes contributing to the nutrient distribution of Colombia River and the Strait of Juan de Fuca. Limnology and Oceanography, 8, 394-410.

Taş, S., Okuş, E., Ünlü, S. & Altıok, H. (2011). A study on phytoplankton following 'Volgoneft-248' oil spill on the north-eastern coast of the Sea of Marmara. *Journal of the Marine Biological Association of the United Kingdom*, 91(3), 715-725.

Tuğrul, S., Beşiktepe, Ş. & Salihoğlu, İ. (2002). Nutrient exchange fluxes between the Aegean and Black Seas through the Marmara Sea, *Mediterranean Marine Science*, 3 (1), 2002, 33-42.

Tuğrul, S., Çoban-Yıldız, Y., Ediger, D. & Yılmaz, A. (2000). Karadeniz, Marmara ve Akdeniz'de partikül organic madde dağılımı ve kompozisyonu, I. Ulusal Deniz Bilimleri Konferansı, 30 Mayıs-2 Haziran 2000, Kültür ve Kongre Merkezi, ODTÜ, Ankara.

Tüfekçi, V., Balkıs, N., Polat-Beken, Ç., Ediger, D. & Mantıkçı, M. (2010). Phytoplankton composition and environmental conditions of a mucilage event in the Sea of Marmara. *Turkish Journal of Biology*, 34, 199-210.

Ullyott, P. & Pektaş, H. (1952). Çanakkale Boğazındaki yıllık temperatür ve tuzluluk değişmeleri hakkında ilk araştırmalar, *Hidrobiologi Mecmuası, İstanbul Üniversitesi Fen Fakültesi Hidrobiyoloji Araştırma Enstitüsü Yayınlarından*, Seri A, cilt 1, sayı, 1, 19-33.

Ünlüata, U., Oğuz, T., Latif, M.A. & Özsoy, E. (1990). On the Physical Oceanography of the Turkish Straits, In: The Physical Oceanography of Sea Straits, L.J. Pratt (Ed.), NATO/ASI Series, 318, 26-60, Kluwer Academic Publishers, Dordrecht.

Vollenweider, R.A., Rinaldi, A., Viviani, R. & Todini, E. (1996). *Assessment of the State of Eutrophication in the Mediterranean Sea. MAP Technical Report Series* 106. UNEP/FAO/WHO., Athens.

Vollenweider, R.A., Giovanardi, F., Montanari, G. & Rinaldi, A. (1998). Characterization of the trophic conditions of marine coastal waters with special reference to the NW Adriatic Sea: Proposal for a trophic scale, turbidity and generalized water quality index. *Environmetrics*, 9, 329-357.

Winkler, L.W. (1888). The determination of dissolved oxygen in water. *Berichte der Deutschen Chemischen Gesellschaft*, 21, 2843-2855.

Yang, X., Wu, X., Hao, H. & He, Z. (2008). Mechanisms and assessment of water eutrophication. *Journal of Zhejiang University Science B*, 9(3), 197-209.

Yılmaz, A. (2002). Biogeochemistry of the seas surrounding Turkey: Cycling and distributions. *Turkish Journal of Engineering and Environmental Sciences*, 26, 219-235.

Yüce, H. & Türker, A. (1991). Marmara Denizi'nin fiziksel oşinografik özellikleri ve Akdeniz suyunun Karadeniz'e girişi, *Uluslararası Çevre Sorunları Sempozyumu Tebliğleri, İstanbul Marmara Rotary Kulübü, İstanbul*, 284-303.

2

Ecological Water Quality and Management at a River Basin Level: A Case Study from River Basin Kosynthos in June 2011

Ch. Ntislidou, A. Basdeki, Ch. Papacharalampou,
K. Albanakis, M. Lazaridou and K. Voudouris*
*Interdisciplinary Postgraduate Study Program
"Ecological Water Quality and Management at a River Basin Level"
Departments of Biology, Geology & Civil Engineering,
Aristotle University of Thessaloniki, Thessaloniki,
Greece*

1. Introduction

The European Parliament and Council decided a policy on the protection, an appropriate treatment and management of water field leading on the Water Framework Directive 2000/60/EC (WFD, European Commission, 2000) in October 2000. The WFD obliges Member States to achieve the objective of at least a good ecological quality status before 2015 and requires them to assess it by using biological elements, supported by hydromorphological and physico-chemical ones. The assessment must be done at a basin level and authorities are obliged to follow efficient monitoring programs in order to design integraded basin management plans. Efforts are being made to adapt national programmes for the WFD requirements (Birk & Hering, 2006). In most European countries, river monitoring programmes are based on benthic macroinvertebrate communities (Sánchez-Montoya et al., 2010).

The WFD (EC, 2000) suggests a hierarchical approach to the identification of surface water bodies (Vincent et al., 2002) and the characterization of water body types is based on regionalization (Cohen et al., 1998). The directive proposes two systems, A and B, for characterizing water bodies according to the different variables considered (EC, 2000). The WFD allows the use of both systems, but considers system A as the reference system. If system B is used by Member States, it must achieve at least the same degree of differentiation. System A considers the following obligatory ranged descriptors: eco-region, altitude, geology and size, whereas system B considers five obligatory descriptors (altitude, latitude, longitude, geology and size) and fifteen optional ones.

A prerequisite for a successful implementation of the WFD in European waters is the intercalibration of the national methods for each biological quality element on which the

*Corresponding Author

classification of ecological status is based (Simboura and Reizopoulou, 2008). According to the Mediterranean intercalibration exercise (MED-GIG) (Casazza et al., 2003, 2004, five river types are proposed, based on the catchment area, the altitude, the geological background and the flow regime of the rivers. Greece participates in this exercise and belongs in the Mediterranean geographical intercalibration group (MED-GIG) (Casazza et al., 2003, 2004).

The pressures and impacts play a key role in the likelihood that a water body will fail to meet the set objectives. IMPRESS analysis (CIS Working Group 2.1: IMPRESS, 2003) assesses the impact and evaluates the likelihood of failing to meet the directive's environmental objectives. Additionally, the Driving force-Pressure-State-Impact-Response (DPSIR) framework represents the relations between socio-economic driving forces and impact on the natural environment (Kristensen, 2004) and the SWOT analysis helps the understanding of the Strength-Weakness-Opportunities-Threats.

This chapter deals with the ecological water quality of the Kosynthos river basin based on (a) the distinction of the water bodies by applying System B and taking into consideration the pressures, (b) the calculation of an approximate water balance according to the activities developed in the river basin, (c) the assessment of the ecological water quality, using benthic macroinvertebrates, (d) the implementation of Impress analysis DPSIR and SWOT analyses.

2. Study area

The Kosynthos River is located in the north-eastern part of Greece, flows through the prefectures of Xanthi and Rhodopi and discharges into the Vistonis lagoon (Figure 1) as a result of the diversion of its lowland part in 1958. Kosynthos' length is approximately 52 Km (Pisinaras et al., 2007). In the present study, 8 sites were selected in Kosynthos river basin (Figure 1) during the period June 2011, depending on the different pressures that presented in the area. Four sites belonged to the mountainous area and the rest sites to the low-land one. The Kosynthos river basin belongs to the water district of Thrace (12th water district), covering an area of 460 Km². The region consists of forest and semi-natural areas (69.6%), rural areas (27.7%), artificial surfaces (2.5%) and wetlands (0.3%) (Corine Land Cover 2000). It is considered to be a mountainous basin (Gikas et al, 2006) of steep slopes and its average elevation is about 702 m. In total, the 7.3% of the basin is protected by the Ramsar Convention or belongs to the EU Natura 2000 sites.

Geologically speaking, the study area belongs entirely to Rhodope massif (Figure 2) consisting of old metamorphic rocks (gneisses, marbles, schists), observed mainly in the northern part of the basin. Moreover, igneous rocks (granites, granodiorites) have intruded the Rhodope massif through magmatic events during Tertiary and outcrop in the central part of the basin. Because of the granite intrusion in the calcareous rocks and the contact metamorphosis, a sulfur deposit is created, consisting mainly of pyrites. Quaternary and Pleistocene mixed sediments cover the south-eastern part of the catchment. The boundary between the highland area and the lowland is characterized by a sharp change of slope.

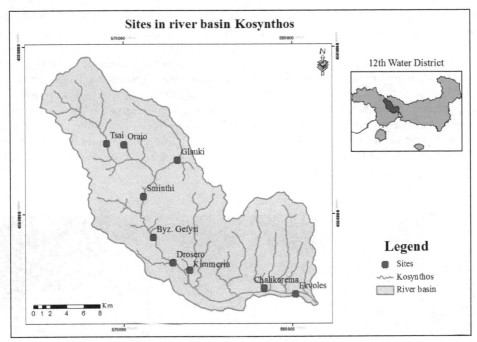

Fig. 1. Map of Kosynthos river basin showing the sampling sites.

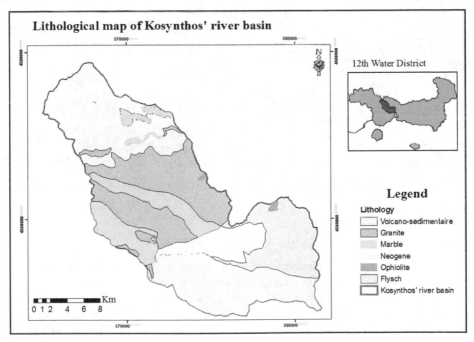

Fig. 2. Lithological map of Kosynthos' river basin.

From a hydrogeological point of view, two main aquifers are developed within the aforementioned geological formations: 1) an unconfined aquifer in the Quaternary deposits of lowlands and 2) a karst aquifer in marbles of the northern part of the basin (Diamantis, 1985). Karst aquifer system often discharges groundwater through springs in the hilly part of the basin, where permeable marbles are in contact with impermeable basement rocks. Previous studies (Hrissanthou et al., 2010; Gikas et al., 2006) show significant sediment transportation to Vistonis lagoon from Kosynthos river because of intense erosion. However, no deltaic deposits are observed in the outfall of Kosynthos, while an inner delta is created right before the stream's diversion (Figure 3). The steep topography combined with the inclination of the diverted section prevents the transportation of coarse sediments, allowing only fine-grained fractions to Vistonis lagoon.

Fig. 3. The inner delta of Kosynthos River, right before the diverted part (Google Earth).

3. Material and methods

3.1 Typology

In this study system B was selected because the basin of Axios River (a transboundary Greek-FYROM river) belongs to two different ecoregions according to System A. In order to distinguish the water bodies of the Kosynthos river basin, apart from the obligatory descriptors the slope, from the optional ones, was selected and a new category in the basin descriptor was added (0-10 Km^2). The rivers were characterized according to the MED-GIG intercalibration exercise (Van de Bund et al, 2009).

3.2 Approximate water balance

The estimation of the approximate water balance of Kosynthos catchment is based on monthly rainfall and temperature data of 7 weather stations (Genisea, Iasmos, Xanthi, Semeli, Gerakas, Thermes, Dimario) distributed equally across and beyond the basin, for the period 1964-1999 and GIS technique (Voudouris, 2007). As part of the estimation process, components of the hydrological cycle (precipitation P, actual evapotranspiration E, infiltration I and surface runoff R), instream flow, available water capacity and water needs (demand for urban, farming, irrigation and industrial water) of the river basin are calculated.

3.3 Quality elements

Dissolved oxygen (DO mg/l), water temperature (WTemp, ºC), pH and conductivity (μS/cm) were measured in situ with probes (EOT 200 W.T.W./Oxygen Electrode, pH-220, CD-4302, respectively). TSS (mg/l), nutrients (N-NH$_4$ and P-PO$_4$, mg/l) and oxygen demand (BOD$_5$, mg/l) were estimated following A.P.H.A. (1985). Flow was quantified with a flow meter (type FP101) and stream discharge (m^3/s) was calculated for each site. The percentage composition of the substrate was visually estimated according to Wentworth (1922) scale. The Habitat Modification Scores (HMS) was calculated to assess the extent of human alterations at each site (Raven et al., 1998).

Benthic macroinvertebrates were collected using a standard pond net (ISO 7828:1985, EN27828:1994) with the semi-quantitative 3-minute kick and sweep method according to Armitage et al (1983) and Wright (2000) proportionally to the approximate coverage of the occurring habitats (Chatzinikolaou et al., 2006). The animals were preserved in 4% formaldehyde.

In the laboratory, they were sorted and identified to family level. To assess the ecological quality of each site the Hellenic Evaluation System (HES) (Artemiadou & Lazaridou, 2005) and the European polymetric index STAR ICMi (European Commission 2008/915/EC) were applied to the benthic macroinvertebrate samples.

3.4 Statistical analysis

For the statistical analyses all data were log (x+1) transformed except for pH and temperature which were standardized. Parameter expressed as percentages (substrate) was arcsine transformed (Zar, 1996). The hierarchical clustering analysis, based on Bray-Curtis index (Clarke and Warwick, 1994) was applied to the samples of benthic macroinvertebrates for grouping them.

Similarity percentages analysis (SIMPER analysis) (Clarke & Warwick, 1994) was used to distinguish the macroinvertebrate taxa contributing to similarity and dissimilarity between the groups. Redundancy Analysis (RDA) was performed in order to detect covariance between environmental variables and abundances of taxa (Ter Braak, 1988). Correlated variables were excluded with the use of the inflation factor (<20) and the Monte Carlo permutations test (p<0.05).

3.5 Impress analysis/DPSIR and SWOT analysis

Impress analysis estimates the impacts taking into account the morphological alterations and the pollution pressures. The morphological alterations estimated through the calculation of a Habitat Modification Score (HMS) (Raven et al., 1998) which is based on the artificial modifications. The pollution pressures are treated differently for point and non point sources. As point sources of pollution are considered the urban wastewater and septic tanks, producing BOD, N and P combinations, which are calculated according to the emission factors (Fribourg-Blanc and Courbet, 2004) whereas livestock according to Ioannou et al., (2009) and Andreadakis et al., (2007) calculated the pollutants.

The human population and the species numbers of breeding animals derived from the Greek National Statistical Service. Industries data, the point sources of pollution, are not available from National Services. Non point sources of pollution, being the land uses, are determined using the Corine Land Cover 2000 and their pollutants are calculated according to the immission factors of WL-Delft et al, (2005). The morphological alterations of pressures were significant if the agricultural land cover was more than 40% (LAWA, 2002) and urban land cover more than 2.5% (Environment Agency, 2005) of the total extent of the river basin.

The pressures from pollution sources would be significant if the total immissions exceeded the proposed limits for irrigation (Decision 4813/98) and for fish life (European Commission 2006/44/EC). All limit scores were adjusted to the river basin, taking into consideration the river flow, estimated as 5.8 m^3/s (Gikas et al., 2006). Multiplied by the estimated river flow, the limit scores were adjusted to the river basin.

The impact assessment, the evaluation of likelihood of failing to meet the environmental objectives and the risk management used the methodology proposed by Castro et al., (2005). Finally, the conceptual model DPSIR (at a river basin level) and SWOT analysis (at the level of Municipalities Mykis and Dimokritos) were applied.

4. Results

4.1 Typology

In accordance with the hierarchical approach, the river flowing in the basin is separated in two main water bodies, due to the canalization of the low-land part of Kosynthos in 1958. Therefore, the diverted part is characterized as heavily modified water body (HMWB), while the rest of the river is characterized as natural water body (NWB). The classification of the river by types leads in 17 types in the catchment area, of which 15 in the drainage network (Figure 4). Finally, the subdivision of a water body of one type into smaller water bodies according to the existing pressures, results in 44 water bodies, from which in 9 sampling of biological, hydromorphological and physico-chemical parameters were executed in June 2011. Based on the European common intercalibration river types (Van de Bund et al., 2009), two types (RM1 and RM2) appear in the river basin.

4.2 Approximate water balance

The climate is semi-humid with water excess and deficiency during winter and summer respectively (Angelopoulos & Moutsiakis, 2011). The annual rainfall (P) is influenced by the elevation (H) of the region (P=0.92H+625, R^2=0.95). The mean annual precipitation in the basin for the period 1964-1999 is 1085.6 mm (Figure 5). Based on Turc method, the coefficient of the actual evapotranspiration was estimated to be 71% of the mean annual precipitation. The remaining amount is allocated to surface runoff (15.3%) and infiltration (13.4%).

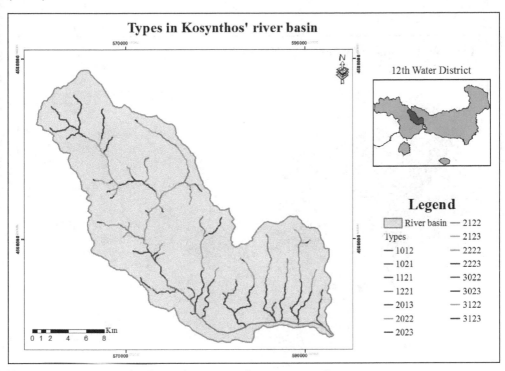

Fig. 4. Types in Kosynthos' river basin according to system B.

A great amount of water infiltrates in marbles and alluvial deposits and then a part of this amount discharged through springs. Instream flow ('environmental flow') is a term that refers to the water required to protect the structure and function of aquatic ecosystems at some agreed level. (Zhang et al, 2006). In accordance with the legislation (M.D. 49828/2008), instream flow equals to 30% of surface runoff and the rest 70% is estimated as available water potential. Assuming that 50% of the infiltration also involves in the available water, the total water potential of Kosynthos basin is calculated to be 86.9x10⁶ m³/yr for the period 1964-1999.

The relevant agents considered for the calculation of water demand are the municipalities that configure the Kosynthos river basin (Municipalities of Myki, Xanthi, Dimokritos, Iasmos). The needs for urban and farming water are calculated equal to 1.7×10^6 m³ each for irrigation water 62×10^6 m³ and for industrial water 6.3×10^6 m³ (demographic and population data 1991-2001. N.S.A.G.). It should be although mentioned that any analysis of water resource management suffers the same handicap with regard to the availability of complete and homogenous information. particularly on municipality level (Torregrosa et al., 2010). Comparing the amount of water potential of Kosynthos catchment for the period 1964-1999 with the water demands, the approximate water balance for the same period is characterized as positive; the water potential is greater that the water demands.

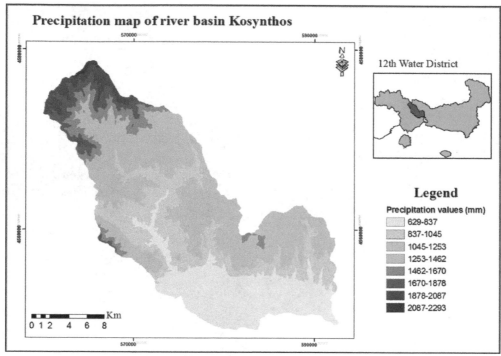

Fig. 5. Precipitation map in Kosynthos river basin.

	P	=	E	+	I	+	R
Water amount (10⁶ m³)	499.4		356.3		66.7		76.4
Precipitation (mm)	1085.6		774		145.5		165.5
Percentage (%)	100		71.3		13.4		15.3

Table 1. Approximate water balance for the Kosynthos river basin (1964-1999).

4.3 Quality elements

The results of the physico-chemical parameters of the river water are presented in Table 2. Ammonium concentration was found to exceed the boundaries of Cyprinid life in all sites, except the site Oraio which exceeded the boundary of portable water and the site Tsai which exceeded boundary of Salmonid life. Also, T.S.S. concentration exceeded the boundary of portable water in sites Kimmeria. Chalikorema and Ekvoles. The substrate composition is represented in Figure 2. The sites Oraio, Byz. Gefyri and Kimmeria are mostly consisted of fine substrate. According to the index HMS most of the sites are characterized as "Predominantly unmodified" (HMS score 3-8, Figure 6).

Sites	D.O. (mg/l)	WTemp (°C)	pH	Conductivity (µS/cm)	T.S.S. (mg/l)	P-PO$_4$ (mg/l)	N-NO$_3$ (mg/l)	N-NH$_4$ (mg/l)	Discharge (m³/sec)	B.O.D.$_5$ (mg/l)
KYA portable water Y2/2600/2001			6.5-9.5	2.5			2.143	11.29	0.318	
Boundaries of Directive 2006/44/EC										
(Salmonid)	6		6.0-9.0		25			0.031		3
(Cyprinid)	4		6.0-9.0		25			0.155		6
Tsai	10.17	15.7	7.52	0.076	1.2	0.022	0.139	**0.141**	0.307	1.97
Oraio	7.29	18.4	7.68	0.312	20.4	0.146	0.715	**0.393**	<0.001	1.45
Glauki	8.38	22.1	8.05	0.258	3.2	0.120	0.424	**0.215**	0.033	1.5
Sminthi	9.22	19.1	7.7	0.17	8.2	0.030	0.309	**0.282**	0.583	1.76
Byz. Gefyri	8.19	26.6	7.71	0.26	4.4	0.041	0.190	**0.207**	0.441	0.94
Drosero	8.68	28.2	8.16	0.25	7.2	0.030	0.190	**0.214**	0.231	0.64
Kimmeria	6.59	33.2	8.01	0.431	36.2	0.024	1.539	**0.228**	<0.001	1.13
Chalikorema	8.25	20.5	6.83	0.501	41.6	0.045	1.248	**0.308**	<0.001	1.33
Ekvoles	8.14	29.1	7.42	0.42	47.6	0.044	0.396	**0.270**	0.306	2.32

Table 2. Physicochemical parameters of the studied sites in the Kosynthos river basin during the period June 2011(with black letters mentioned the concentrations which exceeded the proposed limits).

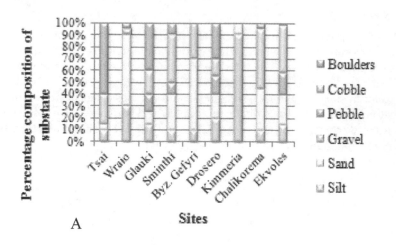

A

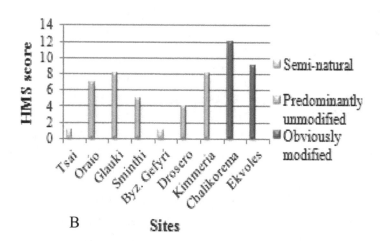

B

Fig. 6. (A) Percentage composition of substrate, and (B) HMS Score at the studied sites of the Kosynthos river basin in period 2011.

In this study 22.005 benthic macroinvertebrates were identified belonging to 48 different taxa. Abundances were found to be higher in the site Oraio and the site Chalikorema had the lowest. The ecological quality of the sites Tsai, Sminthi and Kimmeria, according to the Hellenic Evaluation Score (HES), was characterized as good, sites Oraio Byz. Gefyri, Chalikorema and Ekvoles as moderate and Glauki and Drosero as poor (Figure 7). By the European polymetric index STAR ICMi it was found the same quality, except for the site Kimmeria which was characterized less than good (Table 3). This difference is related to the fact that the HES index takes into account more sensitive taxa.

Sites	Type	Interpretation of HES	Interpretation of STAR ICMi
Tsai	R-M1	Good	Good
Oraio	R-M1	Moderate	< Good
Glauki	R-M1	Poor	< Good
Sminthi	R-M2	Good	Good
Byz. Gefyri	R-M2	Moderate	< Good
Drosero	R-M2	Poor	< Good
Kimmeria	R-M2	Good	< Good
Chalikorema	R-M2	Moderate	< Good
Ekvoles	R-M2	Moderate	< Good

Table 3. The ecological water quality of the studied sites at the river basin Kosynthos in June 2011.

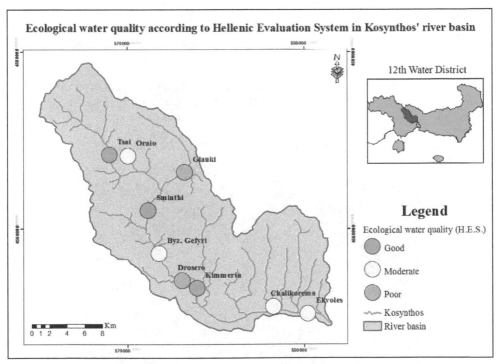

Fig. 7. Ecological water quality according to Hellenic Evaluation System of the studied sites at the river basin Kosynthos in June 2011.

4.4 Statistical analysis

The hierarchical clustering analysis, based on Bray-Curtis index, clustered the benthic macroinvertebrates of the different sites into three clusters (Figure 8). The groups clustered modified sites with an excess of human activities (Ekvoles & Xalikorema) (Group A), the inland delta sites (Drosero & Kimmeria) (Group B) and the high altitude sites (the rest of the stations) (Group C).

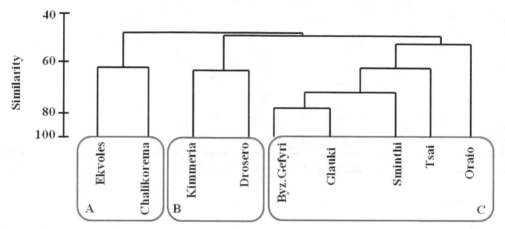

Fig. 8. Hierarchical clustering analysis, based on Bray-Curtis index of the studied sites at the Kosynthos river basin in June 2011.

Simper Analysis showed that the dissimilarity between the groups was around 51%. The families Gammaridae and Simulidae were the key taxa for the differences between the clusters (Figure 9). According to CCA the eigenvalues of the first two axes accounted for 73.8% of the variance. P-PO$_4$ was the variable best correlated with the first axis, whereas the second axis was best correlated with discharge (Figure 10).

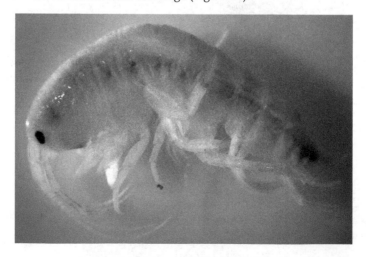

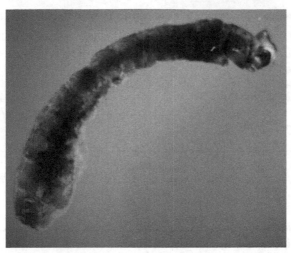

Fig. 9. Benthic macroinvertebrates (A) Gammaridae and (B) Simulidaae which are
responsible for the differences between the groups (Photos: Patsia, 2009).

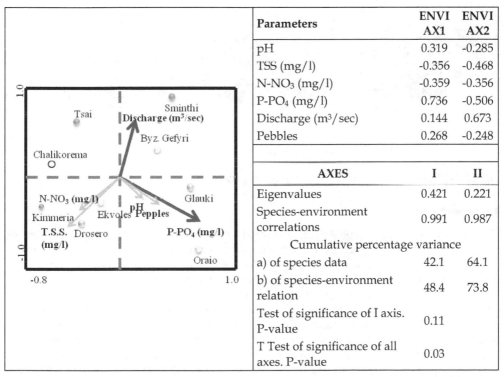

Parameters	ENVI AX1	ENVI AX2
pH	0.319	-0.285
TSS (mg/l)	-0.356	-0.468
N-NO$_3$ (mg/l)	-0.359	-0.356
P-PO$_4$ (mg/l)	0.736	-0.506
Discharge (m^3/sec)	0.144	0.673
Pebbles	0.268	-0.248

AXES	I	II
Eigenvalues	0.421	0.221
Species-environment correlations	0.991	0.987
Cumulative percentage variance		
a) of species data	42.1	64.1
b) of species-environment relation	48.4	73.8
Test of significance of I axis. P-value	0.11	
T Test of significance of all axes. P-value	0.03	

Fig. 10. Canonical correspondence analysis diagram with environmental variables and nine
(9) sites at Kosynthos river basin in June 2011.

4.5 Pressures from pollution sources and morphological alteration pressures

The total emissions, immissions loads produced within the river basin Kosynthos and the environmental quality standards for irrigation and fish life are presented in Table 4. Only BOD exceeded the limits for the salmonid life standard. It is evident that livestock breeding is the most polluting activity (Figure 11). Agriculture is the second diffuse pollution source of total nitrogen (30%) and nitrogen immissions. The morphological alterations for the urban land cover were 1.5% and the agricultural land cover was 27.7% lower than the proposed levels so not significant.

	Emissions	Immissions	Irrigation standards	Fish life standards Salmonid	Cyprinid
BOD (Kg/day)	11.142	2.744	12.562	1.507	3.500
Total N (Kg/day)	4.051	1.814	25.375	2.900	3.200
Total P (Kg/day)	538	85	-	100	201

Table 4. Comparison of emission and immission loads with maximum permitted immission loads.

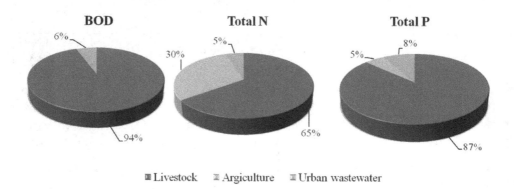

Fig. 11. BOD, total P and total N immissions that each activity produces.

4.6 Impact assessment

The impacts from the morphological alterations are probable, because the mean score of HMS is 5.5. Also, the impacts from the pollution pressures are probable, because the mean biological quality is inferior to good quality and because the nitrogen of N-NH$_4$ exceeds the limit for potable water in the site Oraio. Hence, the impacts from the morphological alterations and pollution pressures are probable. The likelihood of failing to meet the environmental objectives for the morphological alterations is medium because the impacts are probable and there are no significant pressures (urban land cover 1.5% and agricultural land cover 27.7%). Additionally, for the pollution pressures, the likelihood of failing to meet the environmental objectives is medium because the impacts are probable and there are no data for significant pressures (lack of the inputs of industrial pollutants). So in the Kosynthos river basin, an operational monitoring of the risk management for both the morphological and pollution pressures is proposed.

4.7 DPSIR and SWOT analysis

According to the DPSIR framework there is a chain of causal links starting with 'driving forces' (D) (human and economic activities) through 'pressures' (P) (emissions, waste) to 'state' (S) (physical, chemical and biological) and 'impacts' (I) on ecosystems, human health and functions, and eventually leading to political 'responses' (R) (prioritization, target setting, indicators). Consequently, all the above were examined in the Kosynthos river basin (Table 5).

D	P	S	I	R
Urban growth	1. Sewage 2. Urban waste 3. Morphological alterations	1. High concentration in N-NH$_4$ organic pollution and medium water quality 2. Reject of urban solid waste 3. Alteration of river channel and bank (bridges, passage of vehicles). HMS score 5.5	1. Degradation of water quality 2. Alteration of natural landscape and degradation of water quality 3. Interruption of continuity of riparian area and degradation of riparian habitat	1. Creation of installation of treatment of sewages 2 & 3. Collection, recycling and environmental sensitization
Farming activity	Untreated forage sewages	High content in N-NH$_4$ N-NO$_2$-, P-PO$_4$ total organic pollution and medium water quality	Degradation of water quality	Creating organized bands & modern livestock units, wastewater treatment and their use as fertilizer.
Rural activity	Use of pesticides and chemical fertilizers	High concentration in N-NO$_2$ and medium water quality	Degradation of water quality	Sensitization of citizens and farmers
Anthropogenic activities	1 Drilling and over-exploitation of aquifer systems 2. Morphological alterations 3. Deviation and regulation of river watercourse with terraces 4. Forest clearing and sand extraction.	1. Falling water table 2. Degradation of riparian vegetation. low QBR scores in the lowlands 3 & 4. Modification of habitat	1. Depletion of aquifer 2, 3 & 4. Alteration of riparian habitats, increased erosion and input of nutrients.	1.Drilling at regions with highest potential, rational water use and water pricing 2, 3 & 4. Re-establishment of riparian vegetation with native species

Table 5. DPSIR analysis in the Kosynthos river basin

For the sustainable development of the study area, SWOT analysis was applied in Municipality Mykis, which is in the mountainous part of the basin and in Municipality Dimokritos, which is in the lowland part of the basin, in order to estimate the Strengths, Weaknesses, Opportunities and Treats. Based on the SWOT analysis, which is a useful tool for local authorities and decision makers (Diamantopoulou & Voudouris, 2008), some recommendations are proposed to maximize the existing opportunities (Table 6) for achievement of good quality status in Kosynhtos river basin.

Strength	Weakness	Strength	Weakness
Protected areas with biodiversity Byzantine city of 6 AD	Use of septic tanks Absence of treatment in industries Incomplete maintenance of irrigation supply network Over-tapping ground water	Protected areas with biodiversity Archaeological, historical & folklore interest Riparian forest	Use of septic tanks Intensive livestock farming Illegal disposal of debris Illegal logging from riparian forest
Opportunities Growth of ecotourism in protected areas Exploitation of traditional buildings	Threats Absence of urban and industrial waste water treatment Incomplete maintenance of irrigation supply network Absence of administrative policy	Opportunities Growth of ecotourism in protected areas and in the riparian forest Exploitation of traditional buildings Investments	Threats Absence of urban waste water treatment Veterinary surgeon units Absence of administrative policy for the riparian forest
A		B	

Table 6. SWOT analysis in Municipalities: (A) Dimokritos and (B) Mykis.

5. Discussion

In this study System B was selected because of the flexibility in the choice of abiotic parameters and better distinction in relation to the animals than the System A (Dodkins et al., 2005). According to Kanli (2009) the descriptor "Altitude" significantly affects the structure of communities of benthic macroinvertebrates in relation to the other descriptors used in the typology. Also, according to Rundle et al. (1993) and Brewin et al. (1995) "Basin size" is the second most important descriptor that affects the structure of biocommunities after the altitude. In this case, the hierarchical clustering analysis, based on Bray-Curtis index, showed that the descriptor of "Altitude" was the most important descriptor for the separation of benthic macroinvertebrates. For Mediterranean types of RM there was no apparent difference between the stations on the distribution of benthic macroinvertebrates (most of them were R-M2).

The approximate water balance for the period 1964-1999 is characterized as positive, since the water potential in the basin is sufficient to meet the needs arising from activities. The intense infiltration due to the karstic marbles of the Rhodope Mass. and the hydraulic conditions developed in the mountainous area by the presence of impermeable formations does not allow high surface runoff. Moreover, the largest city in the basin (Xanthi) is not watered from this basin.

The concentration of total suspended solids is affected by the dissolution of mineral matter and the intense evaporation (Voudouris, 2009). In this study, the highest TSS concentration measured in the lowland sites (47.6 mg/l) due to the large sediment transportation, mainly fine-grained material derived from the intense erosion, weathering and dissolution of lithological formations because of steep slopes.

The physico-chemical and biological characteristics are modified from the discharge and are related to the ability dissolution of pollutants (Prat et al., 2002). According to Hubbard et al. (2011) the importance of intense flooding in rivers demonstrates the inverse relationship between supply and nutrient concentration. In this study, in the site Kimmeria was found the lowest discharge (0.6 l/s) and low concentration of P-PO$_4$ and N-NH$_4$. This occurs because the actual band width is greater than the measured during the sampling period. Instead the highest concentration of N-NO$_3$ may be due to the influx of water from underground sources upstream of the site. Finally, in the site Oraio it was measured the second smallest discharge (0.8 l/s) and the highest values of nutrients, because the active band width is small and leads to accumulation of nutrients.

The ecological water quality of the site Tsai is connected to the absence of pressures. In the sites Oraio, Glauki and Byz. Gefyri, the ecological water quality is characterized as poor due to the present livestock feeding and the septic tanks. Also, the sites Chalikorema and Ekvoles were characterized as moderate because of the intensive agricultural land use, livestock feeding and septic tanks. Finally, the ecological quality in the site Sminthi and Chalikorema is good, because of the self-purification of the system and the presence of water sources respectively.

Impress Analysis showed that the immissions loads in the basin of Kosynthos is lower than the proposed irrigation limits (Decision 4813/98) issued for another region. It is suggested that an adoption of a similar Decision for Xanthi is important, since it is a rural and agricultural basin with intense activity in the lowland section. Also, the immissions loads did not exceed the limits for the cyprinid life standard, although the total organic load exceeded the limits for salmonid life standard. As livestock breeding appears to be the most polluting activity there is a certain amount of uncertainty involved due to lack of data concerning the location of breeding farms, their grazing fields, their antipollution technologies and the disposal processes of pollutants into the environment (Ioannou, 2009). The latter is mainly due to intense livestock activity observed in the municipality Dimokritos (40% total organic load from the entire river basin). Consequently, a risk management operational monitoring is proposed for both morphological and pollution pressures in the Kosynthos river basin, in order to achieve good quality status in 2015.

6. Conclusions

In conclusion, in the Kosynthos river basin 15 river types are present in the hydrographic network, according to the System B of the WFD. When taking into account the existing pressures in the basin, 44 water bodies are detected. The approximate water balance for the period 1964-1999 is characterized as positive. Among the nine stations selected for sampling benthic macroinvertebrates and according to Hellenic evaluation system of the ecological quality in three stations (Tsai, Sminthi, Kimmeria) water quality was estimated as good, in four stations (Oraio, Gefyri, Chalikorema, Ekvoles) medium and in two (Glauki, Drosero) as poor. Finally, by applying the Impress Analysis, operational monitoring was recommended.

7. References

A.P.H.A. (1985). 'Standard Methods for the Examination of Water and Wastewater'. 16th edition American Public Association.

Andreadakis A., E. Gavalakis, L. Kaliakatsos, C. Noutsopoulos and A. Tzimas (2007). 'The implementation of the Water Framework Directive (WFD) at the river basin of Anthemountas with emphasis on the pressures and impacts analysis'. Proc. Int. Conf. Conference 9th Environmental Science and Technology. Rhodes. Greece.

Angelopoulos C. and E. Moutsikis (2011). 'Hydrological conditions of the Kotyli springs (N. Greece) based on geological and hydrochemical data'. 9th International Hydrogeological Congress. Kalavryta, Greece. 2011.

Armitage P.D., D. Moss, J.F. Wright and M. T. Furse (1983). 'The performance of a new biological water quality score system based on macroinvertebrates over a wide range of unpolluted running-water sites'. Water Research, Vol. 17, pp. 333–347.

Artemiadou V. and M. Lazaridou (2005). 'Evaluation Score and Interpretation Index for the ecological quality of running waters in Central and Northern Hellas'. Environmental Monitoring and Assessment, Vol. 110, pp. 1–40.

Brewin P., T.M.L Newman and S.J. Ormerod (1995). 'Patterns of macroinvertebrate distribution in relation to altitude, habitat structure and land use in streams of the Nepalese Himalaya'. Arch. Hydrobiol., Vol. 135, pp. 79-100.

Birk S. and D. Hering (2006). 'Direct comparison of assessment methods using benthic macroinvertebrates: a contribution to the EU Water Framework Directive intercalibration exercise'. Hydrobiologia, Vol. 566, pp. 401-415.

Casazza G., C. Lopez y Royo and C Silvestri (2004). 'Implementation of the Water Framework Directive for coastal waters in the Mediterranean ecoregion: the importance of biological elements and of ecoregional co-shared application'. Biologia Marina Mediterranea, Vol. 11, pp. 12-24.

Casazza G., C. Silvestri and E. Spada (2003). 'Implementation of the Water Framework Directive for coastal waters in the Mediterranean eco-region'. 6th International Conference on Mediterranean Coastal Environment (MEDCOAST) Eds Ozhan. MEDCOAST Publication, pp. 1157-1168

Castro C.D., AP Infante, J.R. Rodriguez, R.S.X. Lerma and F.J.S. Martinez (2005). 'Manual para la identificacion de las presiones y analisis del impacto en aguas superficiales'. Centro de Publicaciones.

Chatzinikolaou Y., V. Dakos and M. Lazaridou (2006). 'Longitudinal impacts of anthropogenic pressures on benthic macroinvertebrate assemblages in a large transboundary Mediterranean river during the low flow period'. Acta hydrochim. Hydrobiologia. Vol. 34. pp. 453-463.

CIS Working group 2.1: IMPRESS (2003). 'Guidance on the Analysis of Pressures and Impacts'. Office for Official Publications of the European Communities.

Clarke K.R. and M.R. Warwick (1994). 'Change in Marine Communities: An Approach to Statistical Analysis and Interpretation'. Plymouth Marine Laboratory.

Cohen P., H. Andrianamahefa and J.G. Wasson (1998). 'Towards a regionalization of aquatic habitat: distribution of mesohabitats at the scale of a large basin'. Regul Rivers Res Manage. Vol. 14. pp. 391-404

Diamantis J. (1985). 'Hydrogeological study of the basin of Vistonida lake.' PhD thesis. Dep. of Civil Engineering, Polytechnic School. Democritus University of Thrace (in Greek).

Diamantopoulou P., K. Voudouris (2008). 'Optimization of water resources management using SWOT analysis: The case of Zakynthos island, Ionian Sea, Greece'. Environmental Geology 54, pp. 197-211.

Decision 4813/98 (1998). 'Determination of usage of surface waters and special terms for waste disposal in natural receivers at Trikala Prefecture' FEK-575B/11-6-1998.

Dodkins I., B. Rippey, T.J. Harrington, C. Bradley, B.N. Chathain, M. Kelly, M. Quinn, McGarrigle, S. Hodge and D. Trigg (2005). 'Developing an optimal river typology for biological elements within the Water Framework Directive'. Water Research, Vol. 39. pp. 3479-3486.

Environment Agency (2005). 'Technical Assessment Method for Morphological Alterations in Rivers' Water Framework Directive Programme – Environment Agency.

European Commission (2000). 'Directive 2000/60/EC of the European Parliament of the Council of 23rd October 2000 establishing a framework for community action in the field of water policy'. Official Journal of the European Communities.

European Commission (2006). 'Directive 2006/44/EC of the European Parliament of the Council of 6th September 2006 on the quality of fresh waters needing protection or improvement in order to support fish life' Official Journal of the European Communities.

European Commission (2008). 'Commission decision of 30 October 2008 establishing. pursuant to Directive 2000/60/EC of the European Parliament and of the Council. the values of the Member State monitoring system classifications as a result of the intercalibration exercise (notified under document number C (2008) 6016) (Text with EEA relevance) (2008/915/EC)' Official Journal of the European Communities.

Fribourg Blanc B. and C. Courbet (2004). 'Calcul des émissions de polluants organiques et de fertilisants dans l'eau. Second lot: collecte et mise en forme des informations de base nécessaires au calcul des émissions urbaines et industrielles répertoriées, dispersées et surfaciques'. Office International de l'Eau.

Gikas G., T. Yiannakopoulou and V.A. Tsihrintzis (2006). 'Modeling of non-point source pollution in a Mediterranean drainage basin'. Environmental Modeling and Assessment, Vol. 11, pp. 219-233.

Hrissanthou V., P. Deplimani and G. Xeidakis (2010). 'Estimate of sediment inflow into Vistonis Lake. Greece'. International Journal of Sediment Research, Vol. 25, pp. 161-174.

Ioannou A., Y. Chatzinikolaou and M. Lazaridou (2009). 'A preliminary pressure –impact analysis applied in the Pinios River Basin (Thessaly. Central Greece)'. Water and Environment Journal, Vol. 23, pp. 200-209.

Kanli L. (2009). 'Comparison of river typology systems in Greece' MSc thesis. School of Biology. Aristotle University of Thessaloniki. Thessaloniki, Greece.

Kristensen P. (2004). 'The DPSIR Framework'. UN Workshop on a comprehensive assessment of the vulnerability of water resources to environmental change in Africa using river basin approach. UNEP Headquarters. Kenya.

LAWA (2002). 'German Guidance Document for the implementation of the EC Water Framework Directive' Länderarbeitsgemeinschaft Wasser. Mainz, Germany

Ministerial Degree 49828/2008 concerning the 'Adoption of Special Framework for Spatial Planning and Sustainable Development for Renewable Energy and the strategic environmental assessment'.

Ministry of Industry Energy and Technology (1987). 'Weather-raingauge stations'. National Department of Water and Natural Resources. Athens.

National Statistical Agency of Greece Demographic and population data 1991-2001.

Patsia A. (2009). 'Photographic key characters of the Hellenic Evaluation System taxa of benthic macroinvertebrates. Case studies: Strymonas and Nestos rivers' MSc thesis. School of Biology. Aristotle University of Thessaloniki. Thessaloniki, Greece.

Petalas C., F. Pliakas, J. Diamantis and A. Kallioras (2004). 'Study of the distribution of precipitation and surface runoff in the Region of Eastern Macedonia and Thrace for the period 1964-1998'. Bulletin of the Greek Geological Company. Proceedings of the 10th International Congress. Thessaloniki.

Pisinaras V., C. Petalas, A. Gemitzi and V.A. Tsihrintzis (2007). 'Water quantity and quality monitoring of Kosynthos River. North-Eastern Greece'. Global NEST Journal, Vol.9, pp. 259-268.

Raven P., N. Holmes, F. Dawson and M. Everard (1998). 'Quality assessment using River Habitat Survey data'. Aquatic conservation: Marine and Freshwater Ecosystems, Vol. 8, pp. 477-499.

Rundle S.D., A. Jenkins and S.J. Ormerod (1993). 'Macroinvertebrate communities in streams in the Himalaya. Nepal'. Freshwater Biology, Vol. 30, pp. 169–180.

Sánchez-Montoya M.M., M.R. Vidal- Abarca and M.L. Suárez (2010). 'Comparing the sensitivity of diverse macroinvertebrate metrics to a multiple stressor gradient in Mediterranean streams and its influence on the assessment of ecological status'. Ecological Indicators, Vol. 10, pp. 896-904.

Simboura N. and S. Reizopoulou (2008). 'An intercalibration of classification metrics of benthic macroinvertebrates in coastal and transitional ecosystems of the Eastern Mediterranean ecoregion (Greece)'. Marine Pollution Bulletin, Vol. 56, pp. 116-126

Ter Braak C. J. F. and P. Šmilauer (1998). 'CANOCO Reference Manual and User's Guide to CANOCO for Windows. Software for Canonical Community Ordination (version 4)'. Centre for Biometry.

Torregrosa T., M. Sevilla, B. Montano and V. López-Vico (2010). 'The integrated Management of Water Resources in Marina Baja (Alicante. Spain). A Simultaneous Equation Model'. Water Resources Management, Vol. 24, pp. 3799-3815.

Van de Bund WJ (2009). 'Water Framework Directive intercalibration technical report. Part 1: rivers'. JRC Scientific and Technical Reports.

Vincent C., H. Heinrich, A. Edwards. K. Nygaard and J. Haythronthwaite (2002). 'Guidance on typology: Reference conditions and classification systems for transitional and coastal waters'. European Commission.

Voudouris K. (2007). 'An application of SWOT analysis and GIS for the optimization of water resources management in Korinthia prefecture. Greece'. Proc. of International Conference "Water resources management: New approaches and technologies". European Water Resources Association (EWRA). Chania, Crete. 14 16 June 2007, pp. 307-315 (Eds Karatzas G., Papadopoulou M., Tsagarakis K.).

Voudouris K. (2009). 'Environmental Hydrogeology' Tziolas Publications. Thessaloniki, Greece (in Greek).

Wentworth C. (1922). 'A scale of grade and class terms for clastic sentiments'. Journal of Geology, Vol. 30, pp. 377–392.

WL-Delft, S.A. ENVECO and D. Argyropoulos (2005). 'Water Framework Directive Summary Report for articles 5 & 6'. Natural Resources and Environment of Cyprus.

Wright J.F. (2000). 'An introduction to RIVPACS'. In Wright J.F.D., D.W. Sutcliffe & M. Furse (Eds). Assessing the Biological Quality of Freshwaters. RIVPACS and Other Techniques. FBA. Ambleside.

Zar J.H. (1996). 'Biostatistical Analysis' 2nd eds. Prentice-Hall International Inc.

Zhang Y., Z. Yang and X. Wang (2006). 'Methodology to determine regional water demand for instream flow and its application in the Yellow River Basin'. Journal of Environmental Sciences, Vol.18. pp. 1031-1039.

3

Macroinvertebrates as Indicators of Water Quality in Running Waters: 10 Years of Research in Rivers with Different Degrees of Anthropogenic Impacts

Cesar João Benetti, Amaia Pérez-Bilbao and Josefina Garrido
University of Vigo
Spain

1. Introduction

The management of running waters is of great importance for the life of our society and one of the challenges to be met by future generations. The sustainable use of water resources for their exploitation in different aspects is essential. Also, the maintenance of good water quality, both sanitary and environmental, is essential, since it depends largely on the conservation of biodiversity (Fernández-Díaz, 2003).

Rivers are ecosystems of great ecological value with a rich fauna that consists of communities with a complex structure and high biological value. However, their special typology makes them fragile and vulnerable to environmental changes, especially those related to disturbances of anthropogenic origin, which often imply irreversible degradation of their biota (Beasley & Kneale, 2003; Dahl et al., 2004).

The vulnerability of these habitats is also evident in relation to the possible effects of climate change. Among the most affected ecosystems are rivers and streams. One of the predictable effects could be that some of these systems will be transformed from permanent to seasonal and some of them will even disappear. In consequence, the biodiversity of many of them will be reduced and their biogeochemical cycles will be altered (Jenkins et al., 1993).

One of the major impacts that affect rivers is the pollution of their waters by both domestic and industrial waste (Benetti & Garrido, 2010). Also agriculture, with intensive use of fertilizers and pesticides, has contributed significantly to eutrophication and contamination of aquatic ecosystems (García-Criado et al., 1999; Paz, 1993). Another important impact on running waters is the deliberate modification of streams by building dams and reservoirs which alter the ecological characteristics of their basins (Richter et al., 1997).

Fluvial ecosystems support very rich and diverse assemblages, with developed adaptations that allow them to prosper in these environments, and which, at the same time, make them very vulnerable to possible alterations in the habitat. In this sense, human activity often causes severe ecological damage to river systems. These disturbances produce alterations in the chemical composition of water and in the structure of the communities of organisms living in this environment (Oller & Goitia, 2005; Smolders et al., 1999, 2003).

Among the fauna of rivers that should be highlighted are macroinvertebrates. This group of great diversity and ecological importance consists of invertebrates of macroscopic size, normally more than 1mm, living permanently or during certain periods of their life cycle linked to the aquatic environment. They include insects, crustaceans, annelids, molluscs, leeches, etc.

Different groups of macroinvertebrates are excellent indicators of human impacts, especially contamination. Most of them have quite narrow ecological requirements and are very useful as bioindicators in determining the characteristics of aquatic environments (Benetti & Garrido, 2010; Fernández-Díaz et al., 2008; Pérez-Bilbao & Garrido, 2009), to identify the segments of a polluted river where self-purification of organic inputs is under process (Chatzinikolaou & Lazaridou 2007).

The Water Framework Directive (WFD 2000/60/EC) establishes common principles to coordinate the efforts of Member States to improve the protection of the European Community aquatic systems, to promote sustainable water use and to protect ecosystems. This directive is intended to prevent the deterioration of all types of water bodies and to ensure that these environments achieve good quality status. The WFD specifies several quality elements necessary for assessing the ecological state of a river. These elements are hydromorphological, physical, chemical and biological. For the latter the composition and abundance of benthic fauna, including invertebrates, are used. The presence/absence of certain taxa defines the quality state of a watercourse. This directive requires that member states of the European Union achieve good quality of all their water bodies by 2015.

So far, the European intercalibration process has produced class boundaries for four out of five types of Mediterranean rivers (R-M1, R-M2, R-M4 and R-M5) (European Commission, 2008) using benthic macroinvertebrates. The officially selected multimetric index for the intercalibration (European Commission, 2008) of the MedGIG rivers is the STAR_ICMi (Buffagni et al., 2006), which is also used by the Central European and Baltic GIG.

Biological monitoring has been established to control water quality. These studies are often based on the sampling of an area and the subsequent analysis of collected specimens that are suitable for monitoring the area and provide information on pollution trends. The structure of the community of one or more of these specimens (Cheimenopoulou et al., 2011) is used in classifying the watercourse ecological quality in a five-class system by using the ecological quality ratio between the observed to the reference conditions or biotic indices/scores.

Among these indices, diversity indices such as Shannon-Wiener or Margalef have been used or indices or scores based on aquatic macroinvertebrates. In Spain it is used the IBMWP score, which uses the presence of taxa and scores for their tolerance to pollution.

The purpose of this chapter is to study macroinvertebrate fauna as bioindicators of water quality in rivers. The questions are if invertebrates are good indicators of water quality in rivers and which are the effects of the impacts of human activities on invertebrate assemblages living in these habitats.

This chapter presents results from studies conducted over 10 years (1998-2008) in 10 rivers in the Autonomous Region of Galicia (North-western Spain) located in areas with different degrees of anthropogenic impacts. The selection of sampling sites was based on land uses near the river banks (woodlands, agriculture, transport system, urban areas, and industrial activities) in connection to some other habitat parameters. Several abiotic variables were also recorded at the same time as fauna was sampled. Benthic macroinvertebrates and their indices were used for the quality assessment. We also analyzed the biotic indices-environment and assemblage composition-environment relationships in order to study responses to structural characteristics of the habitat (natural or artificial) and variations in water quality parameters.

Fig. 1. Site LI3 of the Limia River (Ourense, NW Spain), before the central power station.

2. Water quality and anthropogenic impacts

Freshwater biodiversity provides a broad variety of valuable goods and services for human societies, some of them irreplaceable (Dudgeon et al., 2006), but human activities have always affected aquatic ecosystems. Rivers are highly vulnerable to change caused by anthropogenic impacts, and their flow is often manipulated to provide water for human use (Bredenhand & Samways, 2009). Globally, the biodiversity of freshwater ecosystems is rapidly deteriorating as a result of human activities (Dahl et al., 2004). According to

Dudgeon et al. (2006), there are five major threat categories to freshwater biodiversity: overexploitation, water pollution, flow modification, destruction or degradation of habitat, and invasion by exotic species.

It is possible that in future decades human pressure on water resources will further endangering aquatic biodiversity present in these systems (Strayer, 2006). Overexploitation of rivers and aquifers for irrigation is already a severe problem in many places, especially in the Mediterranean region. In most countries in the south of Europe, irrigation accounts for over 60% of water (Abellán et al., 2006). This activity can lead to drought and the disappearance of inland aquatic habitats (Abellán et al., 2006; Belmar et al., 2010) or changes in physical and chemical characteristics (Velasco et al., 2006).

Contamination due to different types of pollutants such as fertilizers, sewage, heavy metals or pesticides, is a serious problem worldwide. Increasing urbanization and industrialization generates different non-point sources of contamination, causing impairment of water quality of rivers (Beasley & Kneale, 2003). Many studies have dealt with the negative effect of different pollutants on aquatic biota, which results in biodiversity loss and poor water quality (Beasley & Kneale, 2003, 2004; Benetti & Garrido, 2010; Fernández-Díaz et al., 2008; Garrido et al., 1998; Harper & Peckarsky, 2005; Hirst et al., 2002; Lytle & Peckarsky, 2001; Smolders et al., 2003; Song et al., 2009).

Dam construction is one of the most important modifications that rivers are subjected to (Belmar et al., 2010). The general effect is the transformation of dynamic patterns into static, relatively stable ones with reduced flows (Baeza et al., 2003; Benejam et al., 2010; Stanford & Ward, 1979). Changes in marginal vegetation and in flow velocity may produce changes in the composition of aquatic assemblages, with the replacement of some species by others due to the destruction of microhabitats and the creation of new ones (Fulan et al., 2010; Lessard & Hayes, 2003; Sarr, 2011).

Widespread introduction and invasion of exotic species constitutes another human impact on freshwaters. This usually causes the extinction of indigenous species by competition or predation and biotic homogenization (Raehl, 2002). There are many examples of exotic species invasions, for instance the Nile perch, the American signal crayfish or the zebra mussel.

3. Biological indicators

In the past, water quality was assessed using only physicochemical parameters, but these variables only reflect punctual pollution. The use of biological indicators is more adequate to detect long-term changes in water quality, since aquatic organisms are adapted to specific environmental conditions. If these conditions change, some organisms can disappear (intolerant) and be replaced by others (tolerant). Therefore, variations in the composition and structure of aquatic organism assemblages in running waters can indicate possible pollution (Alba-Tercedor, 1996).

Biomonitoring is the use of biological variables to survey the environment (Gerhardt, 2000). The first step in this type of monitoring is to find the ideal bioindicator whose presence, abundance and behavior reflects the effect of a stressor on biota (Bonada et al., 2006b).

Benthic macroinvertebrates are considered as good indicators of local scale conditions (Metcalfe, 1989; Freund & Petty, 2007). These invertebrates live on the bottom of aquatic ecosystems at least part of their life cycle and can be collected using aquatic nets of 500 µm or less (Hauer & Resh, 1996) or ISO 7828 (EN 27828, 1994). They include molluscs, crustaceans, leeches, worms, flatworms and insects (especially larval stages of some orders). Aquatic macroinvertebrates are used to bioassess aquatic ecosystem quality due to their great diversity of form and habits (Rosenberg & Resh, 1993). According to Johnson et al. (1993) a biological indicator has to fulfil different characteristics:

- to be taxonomically easy and well-known,
- to be widely distributed,
- to be abundant and easy to capture,
- to present low genetic and ecological variability,
- to be preferably big,
- to have low mobility and a long life cycle,
- to present well-known ecological characteristics,
- to have the possibility of being used in laboratory studies

Different sampling protocols and metrics are used to evaluate the water quality of rivers and streams. Among them, biotic indices are the most used because they are highly robust, sensitive, cost-effective, easy to apply, and easy to interpret (Bonada et al., 2006a; Chessman et al., 1997; Dallas, 1995, 1997). Biotic indices are tools for assessing quality based on the different response of organisms to environmental changes (Ministry of Environment, 2005). There are many biotic indices developed for different regions, for instance the TBI (Trent Biotic Index, Woodiwis, 1964), the BMWP (Biological Monitoring Working Party) and the ASPT (Average Score per Taxon) (Armitage et al., 1983; National Water Council, 1981) for the UK, the BBI - Belgian Biotic Index (De Pauw & Vanhooren, 1983; Gabriels et al., 2005) for Belgian rivers, the FBILL (Foix, Besòs i Llobregat, Prat et al., 1999) and the IBMWP (Alba-Tercedor et al., 2002) for Spain, or the HES (Hellenic Evaluation System) (Artemiadou & Lazaridou, 2005) for Greece. Many of these indices have to be adapted when they are used in different regions from where they were developed.

One of the most used biotic indices in the Iberian Peninsula is the IBMWP (Iberian Biological Monitoring Working Party) formerly BMWP' (Alba-Tercedor & Sánchez-Ortega, 1988), which is an adaptation of the British BMWP (Armitage et al., 1983) for the Iberian Peninsula. The taxonomic resolution of this index is mostly at family level, and in some cases is even considered at a higher level. Each benthic macroinvertebrate family (or higher taxa) has a score in relation to their tolerance to pollution, so the sum of the scores of the different taxa found in one site gives a total score allowing this sampling site to be classified in one of the five water quality classes (Alba-Tercedor et al., 2002):

- Class I: very good (≥ 101)
- Class II: good (61-100)
- Class III: acceptable (36-60)
- Class IV: poor (15-35)
- Class V: bad (< 15)

Fig. 2. Central power station of the DevaPO River (Pontevedra, NW Spain).

4. Case study

4.1 Introduction

The negative influence of human impacts, especially pollution, on macroinvertebrate fauna has been described in different studies (Blasco et al., 2000; Dahl et al., 2004; Elbaz-Poulichet et al., 1999; Nummelin et al., 2007; Smolders et al., 2003; Yoshimura, 2008; Artemiadou et al., 2008). In Spain, in recent years such studies have increased considerably and several papers studying these types of impacts in different regions of the country have been published. Amongst them we highlight Alonso (2006) in Madrid, Bonada et al. (2000) and Ortiz et al. (2005) in Catalonia, García Criado & Fernández Aláez (2001) in León, or Marqués et al. (2003) in the Basque Country.

Despite its importance, few studies describe the effects of the impact of hydroelectric power stations, for instance Bredenhand & Samways (2009) in South Africa, Kubecka et al. (1997) in the Czech Republic, Jesus et al. (2004) in Portugal, Lessard & Hayes (2003) in Michigan (USA), Stanley et al. (2002) in Wisconsin (USA), Thomson et al. (2005) in Pennsylvania (E.E.U.U.) or Tonkin et al. (2009) in New Zealand. In Spain we must highlight the study by Oscoz et al. (2006) in Navarre, north of the country, which explored both the impacts of pollution and the impacts of hydroelectric power stations.

So far, in Galicia there are few studies to assess human impact on invertebrate fauna. Some of them have analyzed the effects of anthropogenic impacts on water beetle fauna, one of the most important groups of invertebrates. These studies focused mainly on the effects of water pollution (Benetti & Garrido, 2010; Benetti et al., 2007; Pérez-Bilbao & Garrido, 2009), while Sarr (2011) explores the impact of small hydroelectric stations. The results of these studies provide a basis for conducting this study, which focuses on the impact on the entire fauna of macroinvertebrates. Additionally, this study is partly based on technical reports of different environmental monitoring programs developed by the research group at the University of Vigo and the Ingenieria y Ciencia Ambiental S.L. Company (Garrido et al., 1999, 2000a, 2000b, 2000c, 2003, 2005).

Concerning the macroinvertebrates this study (a) assesses the importance of invertebrate fauna as an indicator of water quality in these rivers; (b) identifies the response of macroinvertebrates to human activities; (c) and denotes the responsible factors for the differentiation of the studied rivers.

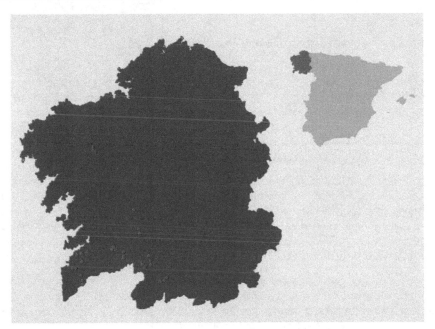

Fig. 3. Map of the study area showing the location of the sampling sites.

4.2 Material and methods

4.2.1 Study area

The study area comprised 10 rivers located in the Autonomous Region of Galicia (North-western Spain) (Figure 3). According to the Köppen-Geiger Climate Classification, the climate of the study area is warm temperate, with dry summers and mild temperature (Kottek et al., 2006). This territory belongs to the Atlantic and the Mediterranean biogeographical regions. Altitude ranges from the sea level to around 1,000 m in the highlands. Due to its geographical location, orography and climate, this area has a large number of running waters, from large rivers to small streams (Figure 1). The landscape consists of woodlands (temperate broadleaf forest, pine and eucalyptus reforestations), farmlands, urban and industrial areas. According to the WFD (Annex II, System A), the studied rivers were classified as: Iberian-macaronesian ecoregion, siliceous/calcareous, lowland/mid-altitude and small/medium catchment area.

These rivers are located in areas with different degrees of anthropogenic impacts. The Lagares River (Figure 4) runs almost entirely through the urban area of Vigo, a city with approximately 300,000 inhabitants. This river has undergone a profound change in its structure, especially channelling, as a result of the growth of the city and rapid industrial development with the consequent establishment of industries on its banks. The rest of the rivers run mostly through rural and semi-urban areas. The course of the rivers Furnia and Miñor has not been altered very much, but the other rivers (Avia, DevaOU, DevaPO, Limia, Tambre, Tea and Tuño) have at least one small hydroelectric power station (Figure 2).

4.2.2 Sampling methods and variables measured

In a period of 10 years, between June 1998 and April 2008, 10 rivers were sampled. Each river was sampled four times, in spring, summer, autumn and winter. In each river 3 sites were selected, located in the upper, middle and low stretches (Table 1). The selection of sampling sites was based on land uses (woodlands, agriculture, transport system, urban areas, and industrial activities) and the position of the hydraulic infrastructures, for the regulated rivers. In these cases the first site was located upstream the dam, the second between the dam and the power station, and the third downstream the station.

Sampling was carried out in all types of substrate for a standardized time (one minute). Fauna was collected with an entomological water net of 30 cm diameter, 60 cm depth and 0.5 mm mesh. The specimens were stored in a 4% formaldehyde solution and taken to the laboratory, where they were sorted and identified. After being studied, they were conserved in 70° alcohol and deposited in the scientific collection of the Aquatic Entomology Laboratory at Vigo University.

The following water quality parameters were measured at each site: temperature, dissolved oxygen, pH, electrical conductivity and total dissolved solids (TDS). Additionally, we measured the altitude (meters above sea level) for each site. These above parameters are considered fundamental in the typology of rivers by the WFD.

4.2.3 Data analysis

The structure of the assemblages was assessed using different diversity indices: richness (S), rarefied richness (ES), abundance (N) and the Shannon-Wiener diversity index (H'). The IBMWP biological index was also calculated. The values of rarefied richness were calculated for 100 individuals ES (100). ES and H' were calculated using PRIMER version 6. Analysis of variance (two-way ANOVA) was used to test for significant differences between seasons in both diversity indices and environmental variables. ANOVA was run using SPSS version 19.

Macroinvertebrates as Indicators of Water Quality in Running Waters: 10 Years of Research in Rivers with Different
Degrees of Anthropogenic Impacts

53

		UTM Coordinate			
River	Site	X	Y	Altitude	Sampling Date
Ávia	AV1	567297	4709038	569	1998-1999
Ávia	AV2	571104	4693658	113	1998-1999
Ávia	AV3	572157	4690485	108	1998-1999
Deva OU	DEOU1	575784	4662895	552	1998-1999
Deva OU	DEOU2	574167	4665277	329	1998-1999
Deva OU	DEOU3	572991	4666511	192	1998-1999
Deva PO	DEPO1	558785	4674976	508	2001-2002
Deva PO	DEPO2	558972	4671670	350	2001-2002
Deva PO	DEPO3	558650	4670559	220	2001-2002
Furnia	FU1	524677	4655852	77	2007-2008
Furnia	FU2	525442	4652027	36	2007-2008
Furnia	FU3	525603	4649199	6	2007-2008
Lagares	LA1	529266	4675118	240	2001-2002
Lagares	LA2	524323	4673627	34	2001-2002
Lagares	LA3	520918	4673068	10	2001-2002
Limia	LI1	593494	4654101	610	2003
Limia	LI2	591737	4652248	600	2003
Limia	LI3	588536	4651147	550	2003
Miñor	MI1	525456	4667235	330	2001-2002
Miñor	MI2	519199	4662061	9	2001-2002
Miñor	MI3	517156	4662213	2	2001-2002
Tambre	TA1	536133	4756871	210	1998-1999
Tambre	TA2	534947	4758037	190	1998-1999
Tambre	TA3	534574	4759534	179	1998-1999
Tea	TE1	550473	4678468	192	1999
Tea	TE2	550950	4677584	114	1999
Tea	TE3	550656	4677503	109	1999
Tuño	TU1	580854	4665249	730	1998-1999
Tuño	TU2	580506	4667324	484	1998-1999
Tuño	TU3	579643	4671803	372	1998-1999

Table 1. Sampling sites, with their code, location in UTM coordinates, altitude and sampling
date.

Relationships between environmental variables with diversity indices and fauna were
determined by a Pearson correlation test. Prior to this, the Kolmogorov-Smirnov test was
used to verify the normal distribution of the data. Variables not following the normal
distribution were logarithmically transformed (log_{10}). These analyses were performed using
SPSS version 19.

Canonical correspondence analysis (CCA) was used to analyze fauna-environment relationships in order to identify environmental factors potentially influencing macroinvertebrate assemblages. A Monte Carlo permutation test was performed to assess the statistical significance of the environmental parameters and the full model to arrive at the significance of the first two canonical axes (Heino, 2000). The environmental factors used were pH, water temperature, dissolved oxygen and conductivity. TDS was not considered because it was redundant with conductivity. CCA was carried out on global abundances, that is, total number of individuals collected at a site over the sampling period. Taxa with less than 10 individuals were removed from the analysis, which was carried out using the CANOCO 4.5 program (Ter Braak & Šmilauer, 2002).

Complete linkage cluster analysis with Bray-Curtis coefficient was used to cluster the rivers into groups and thus be able to verify differences in assemblage composition. The results were represented graphically by Multidimensional scaling (MDS). SIMPER analysis was used to identify which species generate the most similarity within each MDS group. For the SIMPER routine, the raw data were square root transformed and reporting was limited to species with more than 2.5% contribution to dissimilarity. This analysis was carried out with PRIMER version 6.

Fig. 4. Lagares River (Pontevedra, NW Spain) impacted by pollution.

Macroinvertebrates as Indicators of Water Quality in Running Waters: 10 Years of Research in Rivers with Different
Degrees of Anthropogenic Impacts

55

4.3 Results

4.3.1 Diversity and biological indices

In total, 115 taxa (mostly family) of 7 phyla were collected (Table 3). The most representative groups were insects (86 families), especially the orders Diptera (20 families), Trichoptera (19 families) and Coleoptera (12 families).

Only the family Elmidae was recorded in all sites. Other very common families were Baetidae and Chironomidae (recorded in 29 sites), and Nemouridae and Simuliidae (recorded in 28 sites). Besides, 10 families (Bithyniidae, Capniidae, Hebridae, Heteroceridae, Libellulidae, Noteridae, Pyralidae, Scatophagidae, Sciomyzidae and Sisyridae) were only recorded in one site.

If we analyse these results by samples, we can see that site 3 in the Limia River (LI3) presented the highest value of invertebrate richness in autumn with 68 taxa recorded, followed by FU2 (Furnia River) in spring and DEOU1 (DevaOU River) in winter with 55 taxa each. On the other hand, the lowest richness was found in LA3 (Lagares River) in spring, where only 7 taxa were collected, followed by TA2 (Tambre River) in winter and in autumn with 8 and 11 taxa recorded respectively.

Total abundance was 217,577 individuals (185,287 insects, 14,931 Annelida, 9,430 Mollusca and 7,929 other groups). The greatest abundance was observed in the third site in the Limia River in spring, with 29,931 individuals collected. On the other hand, the lowest value was obtained in the third site of the Lagares River in spring with 59 individuals. The most abundant family of macroinvertebrates was Chironomidae with 40,584 individuals recorded. Other abundant families were Ephemerellidae (25,788 individuals), Bactidae (22,860), Elmidae (15,171) and Simuliidae (13,611).

Rarefied species richness is the expected number of species for a given number of randomly sampled individuals and facilitates comparison of areas in which densities may differ (McCabe & Gotelli, 2000). The highest values correspond to the Limia River, sites LI3 (24.18) and LI2 (24.13) in autumn, and to the Furnia River, site FU2 in autumn (23.51) and spring (23.27). On the contrary, the Lagares River had the lowest values in site LA2 (6.62) and site LA3 (6.92).

The Shannon-Wiener index (H'(log2)) revealed that most of the studied rivers presented high diversity values. The lowest diversity was recorded in DEPO3 (DevaPO River) in winter (0.86) and the highest in LI3 (Limia River) in autumn (4.40). In general, the diversity values were high, greater than 3 in 60% of the samples.

According to the IBMWP index, most samples (95%) presented good water quality (> 60, class II), even 87% of the samples presented very good quality, because the index values were above 100 (class I). The highest value (338) was obtained at site FU2 in the Furnia River and the lowest (26) at site LA3 in the Lagares River. Only 2 samples, belonging to the Lagares and Tambre rivers, obtained low values, below 35 and therefore classified as poor quality (class IV). No samples presented bad water quality (< 15, class V).

Table 2 shows the mean minimum and maximum values of the diversity indices for the studied rivers. There were no significant differences (p < 0.05) among seasons in any of the diversity indices, as evidenced by the ANOVA.

Richness measures	Mean ± SD	Minimum	Maximum	ANOVA	
				F	p
Richness S	30.98 ± 11.84	7	68	0.128	0.943
Rarefied Richness ES (100)	15.70 ± 4.62	6.62	24.18	0.502	0.682
Abundance N	2175.77 ± 3445.50	59	23931	1.579	0.202
Diversity H'(log2)	3.04 ± 0.75	0.86	4.40	0.535	0.660
IBMWP	174.51 ± 70.47	26	338	0.181	0.909

Table 2. Mean, SD and ranges of biological and diversity indices of the samples and ANOVA with season as factor.

4.3.2 Environmental variables

Table 4 shows mean minimum and maximum values of the environmental variables measured in the 10 studied rivers. The main result to highlight is the high value of conductivity measured in some sites, especially in the Lagares River, higher than in the other surveys. ANOVA showed no significant differences ($p < 0.05$) among seasons in almost all environmental variables. This analysis only showed significant differences among seasons in temperature, as expected.

4.3.3 Influence of environmental variables on macroinvertebrate assemblages

The Pearson correlation test was performed to assess the relation between the environmental variables and the taxa and diversity indices. We found several significant correlations ($p < 0.05$), but most of them were low ($r < 0.5$). We only highlight those that were higher ($r > 0.5$). Regarding the diversity indices, there were significant negative correlations between conductivity and rarefied richness ($r = -0.52$) and diversity ($r = -0.50$). For the taxa we found a significant negative correlation between oxygen concentration and Naididae ($r = -0.81$), and a significant positive correlation between conductivity and Hydrobiidae ($r = 0.51$).

Figure 5 shows the results of the CCA. The eigenvalues for axes 1-4 were 0.400, 0.198, 0.092 and 0.074 respectively. Correlations for axes III and IV with environmental variables were low ($r < 0.5$), and only axes I and II were used for data interpretation. The cumulative percentage of variance for the species-environmental relation for these two axes was 76.1%. The first two canonical axes were significant, as shown by the Monte Carlo permutation test ($p = 0.004$).

The first principal axis is positively correlated with conductivity ($r = 0.892$) and temperature ($r = 0.788$), and negatively with altitude ($r = -0.532$). This component describes water quality and can be an indicator of contamination, since all Lagares river sites are located at the positive end of the axis. The second axis is positively correlated with temperature ($r = 0.396$) and negatively with oxygen ($r = -0.786$). Rivers Limia and Miñor are located at the negative end of this axis, indicating low oxygen values. According to the CCA analysis, the family Naididae is related to sites with low oxygen values and the family Hydrobiidae to sites with high conductivity values, while Enchytraeidae prefer sites with high pH values.

Macroinvertebrates as Indicators of Water Quality in Running Waters: 10 Years of Research in Rivers with Different Degrees of Anthropogenic Impacts

57

Higher taxa	Taxa	Ávia	Deva OU	Deva PO	Furnia	Lagares	Limia	Miñor	Tambre	Tea	Tuño
Hydrozoa	Hydridae		5			1					
Turbellaria	Dugesiidae	16				2	159				
Turbellaria	Planariidae	37	366	256	72	43	14	27			14
Nematoda	Nematoda		11		2	82	53	3			5
Nematomorpha	Gordiidae		2								1
Hirudinea	Erpobdellidae	55	1	16	2	20	84	2	30		
Hirudinea	Glossiphoniidae	10	32			1	15		8	17	753
Hirudinea	Hirudinidae		6		1						
Oligochaeta	Enchytraeidae	1621	735	595		1957	1386	45	726	127	413
Oligochaeta	Lumbricidae	65		162			53		25		
Oligochaeta	Lumbriculidae		25			27	320			3	40
Oligochaeta	Naididae	1	7	492		646	2609	1261	25	220	
Oligochaeta	Oligochaeta unidentified				61						
Oligochaeta	Proppapidae						62				
Oligochaeta	Tubificidae	84		13			72				
Bivalvia	Sphaeriidae	10	23	1	1	11	193	9	104		11
Gastropoda	Ancylidae	20	620	11	10	15	282		33	8	111
Gastropoda	Bithyniidae		1								
Gastropoda	Hydrobiidae	6	84			6960		15	218	2	2
Gastropoda	Lymnaeidae	79	8		13		119		148		5
Gastropoda	Physidae	6				39	213		32		
Gastropoda	Planorbidae	1	1		1		1				
Gastropoda	Valvatidae		2						1		
Crustacea	Cladocera					1	45				
Crustacea	Copepoda		12			10	184				
Crustacea	Ostracoda				1	31		1			
Crustacea	Asellidae		17				233				5
Crustacea	Gammaridae	2	8	11	534	37		70	805		421
Arachnida	Hydracarina	66	1244	188	34	83	2311	65	130	187	22
Collembola	Collembola				8	20	163	3			
Odonata	Aeshnidae	24	64	17	14	2	17		44	22	28
Odonata	Calopterygidae	18	47	26	55	2	21		60	7	17
Odonata	Coenagrionidae	2					14				
Odonata	Cordulegasteridae	12	31	5	5	4	92	3	1	11	47
Odonata	Gomphidae	102	6	3	57	1	709	13	40	54	1
Odonata	Lestidae				2		20				
Odonata	Libellulidae						2				
Plecoptera	Capniidae						1				
Plecoptera	Chloroperlidae	133	7	131	41		6	32	1	6	1
Plecoptera	Leuctridae	430	623	161	272		1789	46	419	73	1458
Plecoptera	Nemouridae	670	871	301	264	11	58	368	30	102	403
Plecoptera	Perlidae	11	48	6	8		2		26	37	50
Plecoptera	Perlodidae	1	2	35	84		2	2		13	33

Order	Family										
Ephemeroptera	Baetidae	551	3765	2638	371	2411	6078	3117	1146	850	1933
Ephemeroptera	Caenidae	401				1	1094		256	4	
Ephemeroptera	Ephemerellidae	213	864	591	278	28	22547	42	578	257	390
Ephemeroptera	Ephemeridae		328	4	6		20			3	87
Ephemeroptera	Heptageniidae	180	933	57	45	35	2909	37	82	27	184
Ephemeroptera	Leptophlebiidae	1004	505	2342	258		119	132	13	112	428
Ephemeroptera	Oligoneuriidae		1				253		9		3
Ephemeroptera	Potamanthidae			532							
Hemiptera	Aphelocheiridae				1		215		112	54	
Hemiptera	Corixidae	6					33		1	4	
Hemiptera	Gerridae	13	16	54	17	25	84	11	76	40	20
Hemiptera	Hebridae						3				
Hemiptera	Hydrometridae		9	5			16		2		
Hemiptera	Mesoveliidae	2	1				1				
Hemiptera	Naucoridae				1		47				
Hemiptera	Nepidae				1		4				
Hemiptera	Notonectidae						4				
Hemiptera	Veliidae				1		2		2		
Coleoptera	Dryopidae	3	13	9	1		48	1	36	11	19
Coleoptera	Dytiscidae	9	14	79			799	1	8		4
Coleoptera	Elmidae	663	1899	3615	448	22	4474	200	1702	298	1850
Coleoptera	Gyrinidae		1	51	73		112		11	9	11
Coleoptera	Haliplidae					2	165				
Coleoptera	Helophoridae			1			6		1		
Coleoptera	Heteroceridae						40				
Coleoptera	Hydraenidae	395	500	396	126	1	414	24	15	33	239
Coleoptera	Hydrochidae		1				44				
Coleoptera	Hydrophilidae	4	1	12	1	1	139	2	23	3	
Coleoptera	Noteridae						1				
Coleoptera	Scirtidae		20	25	62		2	1		48	2
Diptera	Anthomyiidae					8	268	1	1	9	
Diptera	Athericidae	132	298	31	35	23		30	200	60	24
Diptera	Blephariceridae		3		3		17				
Diptera	Ceratopogonidae	28	106	189	21	41	2	9	10	8	27
Diptera	Chironomidae	1597	9418	13729	1326	3575	376	5287	1750	620	2906
Diptera	Culicidae						7471	1			
Diptera	Dixidae	6	30	2	5		35			3	4
Diptera	Dolichopodidae	8	123				2		5	6	5
Diptera	Empididae	22	348	221	208	94	1	74	13	81	133
Diptera	Limoniidae	62	57	18	19		1005	11		10	66
Diptera	Psychodidae	11	31	402	13	39	4	13		29	56
Diptera	Ptychopteridae						21				
Diptera	Rhagionidae	3	7	3	3	2	26				1

Macroinvertebrates as Indicators of Water Quality in Running Waters: 10 Years of Research in Rivers with Different
Degrees of Anthropogenic Impacts

59

Diptera	Scatophagidae				2						
Diptera	Sciomyzidae	11									
Diptera	Simuliidae	787	3832	2829	1020	550	2	2740	559	168	1124
Diptera	Stratiomyidae						51				
Diptera	Tabanidae	2	7	2	1		762	3		9	2
Diptera	Thaumaleidae						3				
Diptera	Tipulidae	20	25	9	1	4	2		11	6	12
Lepidoptera	Pyralidae						9				
Megaloptera	Sialidae	29	22	2			2		1		
Neuroptera	Sisyridae						37				
Trichoptera	Beraeidae	5	4					3	19	13	1
Trichoptera	Brachycentridae	18		1496	34		651	6	21		10
Trichoptera	Calamoceratidae	5	24		5		64		175		25
Trichoptera	Ecnomidae		2				4				6
Trichoptera	Glossosomatidae	1	63	8	34	1	1	3			3
Trichoptera	Goeridae	5	148	7	1					1	1
Trichoptera	Helicopsychidae	43	107	15	24	1	1	3			17
Trichoptera	Hydropsychidae	291	837	567	466	82	4367	43	482	118	199
Trichoptera	Hydroptilidae	2		9	14		38	1			
Trichoptera	Lepidostomatidae	25	43	40	111	3		41	1	22	2
Trichoptera	Leptoceridae	49	69	71	12	5	20	3		9	5
Trichoptera	Limnephilidae	39	76	172	7		4416	1	3	5	12
Trichoptera	Philopotamidae	61	98	179	569	4	5421	63		22	9
Trichoptera	Phryganeidae		4				32				
Trichoptera	Polycentropodidae	27	45	94	11		663	4	19	5	53
Trichoptera	Psychomyiidae	5	5	73	5	1	10				
Trichoptera	Rhyacophilidae	59	235	147	56	20	200	33	58	14	124
Trichoptera	Sericostomatidae	59	463	222	124	1	8	3	2	67	173
Trichoptera	Uenoidae	20	16	2	19			20			1

Table 3. Abundance of macroinvertebrates recorded in 10 rivers between 1998 and 2008.

				ANOVA	
Variable	Mean ± SD	Minimum	Maximum	F	p
Water Temperature °C	13.90 ± 3.38	7.90	22.00	70.536	0.000
pH	6.43 ± 0.63	5.20	8.17	0.297	0.828
Conductivity µS/cm	60.95 ± 38.12	20.00	186.00	0.975	0.410
Oxygen %	97.83 ± 5.76	80.40	107.00	1.436	0.239

Table 4. Mean, SD, minimum and maximum values of the environmental variables
measured in the studied rivers and ANOVA with season as factor.

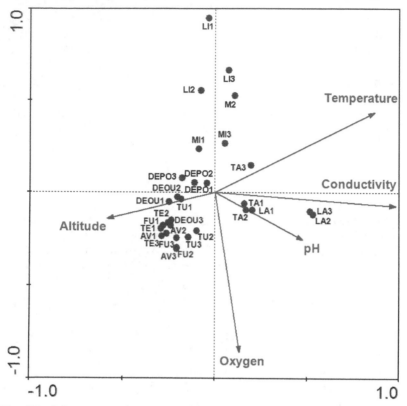

Fig. 5. Results of the canonical correspondence analysis (CCA) based on the invertebrates assemblages with respect to environmental variables. Arrows represent the environmental variables and circles the sites.

4.3.4 Assemblage composition

The Bray-Curtis coefficient was used to calculate the affinity between rivers and the results ranged from 0 to 100. If the value obtained is close to 100, populations will be more similar. According to this coefficient the studied rivers had a 36.32 average faunal affinity. The greatest degree of affinity was observed between the rivers DevaOU and DevaPO (68.20), and between the Tambre and Tuño (61.95). On the other hand, the rivers faunistically farthest were the Limia and Tea, with 7.66 of affinity. MDS analysis provided alternative insights into the similarity of sites with regard to macroinvertebrate assemblage composition. Figure 6 shows the formation of four clearly separated groups with low faunal affinity between them, less than 40%.

The Limia River has an affinity of less than 25% with all the other rivers, forming a group apart and completely separated from the rest. The Lagares River also forms a separate group, with a greater affinity with the Miñor River. Finally, we identified two other groups, one formed by the rivers Miñor, DevaPO and DevaOU, and another formed by the rivers Avia, Furnia, Tambre, Tea and Tuño.

Table 5 lists the species that contributed the greatest amount of similarity in groups A and B. We found a relatively high level of overall similarity in group A (59.26%) and B (50.81%) with the SIMPER analysis. The taxa that contributed most to similarity were the same in both groups. The most important taxon was Chironomidae, contributing with 42.64% similarity in the group A and 23.73% similarity in the group B. Other important taxa were Baetidae, Elmidae and Simuliidae.

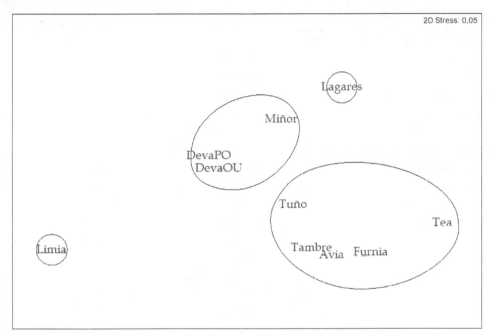

Fig. 6. Multidimensional scaling (MDS) from complete linkage clustering based on the Bray-Curtis coefficient.

4.4 Discussion

The importance of using macroinvertebrates as bioindicators of the water quality of rivers has already been highlighted by several authors (Alba-Tercedor et al., 2002; Alonso, 2006; Bonada et al., 2000; Ortiz et al., 2005; Oscoz et al., 2006). The importance of this group is also reflected in this study because we were able to evaluate the conservation status of these rivers and assess the degree of impact that they are subjected to, whether by pollution or the construction of hydroelectric power stations.

In general, the biological and diversity values observed in the studied sites were high. However, in some places, especially in the Lagares River, these indices values are considerably low in comparison with other rivers in northern Spain (Álvarez-Troncoso, 2004; Fernández-Díaz, 2003; García-Criado, 1999; Paz, 1993; Pérez-Bilbao & Garrido, 2009). We also found a negative correlation between conductivity and richness parameters with a significant decrease in rarefied richness and diversity in sites with high values of this variable, especially in the Lagares River. The reduction in macroinvertebrate richness and abundance in stretches with high values of variables indicative of pollution reflect anthropogenic impacts and corroborates the data obtained by the biological index, which showed poor quality. In relation to this, the results obtained in this study are in concordance with others reported in other papers, which have often documented a decrease in richness and diversity in water bodies with high values of chemical variables (Heino, 2000, Prenda & Gallardo-Mayenco, 1996; Thorne & Williams, 1997).

Taxa	Mean Abundance	Contribution to similarity (%)
Group A (Average similarity: 59.26)		
Chironomidae	9478	42.64
Baetidae	3173.33	18.86
Simuliidae	3133.67	18.48
Elmidae	1904.67	4.34
Group B (Average similarity: 50.81)		
Chironomidae	1639.8	23.73
Baetidae	970.2	13.15
Elmidae	992.2	11.24
Simuliidae	731.6	9.99
Ephemerellidae	343.2	5.74
Leuctridae	530.4	4.7
Hydropsychidae	311.2	4.57
Enchytraeidae	577.4	3.7
Nemouridae	293.8	2.84
Leptophlebiidae	363	2.79
Gammaridae	352.4	2.66

Table 5. SIMPER analysis of macroinvertebrates assemblages of groups A (Miñor, DevaOU and DevaPo rivers) and B (Avia, Furnia, Tambre, Tea and Tuño rivers).

IBMWP index values obtained in the samples of this study, mainly above 100, highlighted the good preservation state of the studied rivers. Besides, it is important to note that the highest value was obtained in the Furnia River, a very little impacted river with excellent water quality. On the contrary, the lowest value was obtained in the Lagares River, which is highly polluted and affected by anthropogenic pressure. The importance of using this index was also pointed out by other studies conducted in the Iberian Peninsula (Alba-Tercedor et al., 2002; Bonada et al., 2006a; Poquet et al., 2009).

In addition to the reduction in richness, abundance and biological indices, pollution also causes a change in faunal composition, as reflected by the separation of the Lagares River from the others in the faunal affinity analysis. In this sense, it is important to note that the dominance of certain taxa, e.g. Naididae and Chironomidae, and absence of others, e.g. Plecoptera, at some sites may indicate the existence of alterations in them (Oscoz et al., 2006).

According to Beasley & Kneale (2003), increasing urbanization and industrialization generates different non-point sources of contamination, causing impairment of water quality of rivers. This environmental impact can be seen in the city of Vigo and its surroundings. High anthropogenic pressure on aquatic ecosystems in this region is a consequence of the ever-increasing population and establishment of industries, especially on the banks of rivers (Benetti & Garrido, 2010).

The response of macroinvertebrates to water pollution seems to define, at least, the species typical of non-contaminated sites. In this sense, ordination analysis identified a group of sensitive taxa especially evident for those most abundant, whose numbers fall considerably in impacted sites. According to CCA, there are also tolerant taxa, for example the Hydrobiidae family, correlated with high values of conductivity, and Naididae, correlated with low values of dissolved oxygen. These results are in agreement with those found in other studies. According to Brinkhurst & Gelder (2001), several aquatic oligochaetes, including Naididae species, have red blood pigments which aid oxygen uptake and transport, thus they can live in environments with low oxygen. Pérez-Quintero (2007) documented that some Hydrobiidae species were salinity-tolerant, so possibly also conductivity-tolerant, as these variables are closely correlated.

Most studies about impacts in rivers (Benetti & Garrido, 2010; Dahl et al., 2004; Nummelin et al., 2007) are mainly focused on the impact assessment of different sources of water pollution, without considering the impacts of infrastructures that change riverbeds, such as the case of small hydro. One of the best ways to evaluate the impact produced by hydroelectric power stations on wildlife is to check changes in the faunal composition and mainly their feeding habits (Argyroudi et al., 2010). In the studied rivers, we observed that the impact produced by hydroelectrics brings about a change in the hydrological regime, mainly the damming, and consequently a change in the community of invertebrates, especially altering its faunal composition, something already noted by other authors for different groups of invertebrates (Bredenhand & Samways, 2009; Jesus et al., 2004; Sarr, 2011; Stanley et al., 2002; Yoshimura, 2008). The structure of macroinvertebrate assemblages is influenced by factors such as the hydrological regime, substrate stability, type and abundance of trophic resources, or land use in the river basin (Dessaix et al., 1995; Quinn et al., 1997; Zamora-Muñoz & Alba-Tercedor, 1996). In our study, both natural characteristics (geology, substrate, water flow) and those artificially created by the impact of hydroelectric infrastructures, determined the structure of the invertebrate assemblages.

The regulation of rivers and hydropower development changes the habitat structure (Dessaix et al., 1995; Dolédec et al., 1996; Fjellheim & Raddum, 1996; Oscoz et al., 2006) and causes the loss of more sensitive taxa and thus imbalance in community structure (Fjellheim et al., 1993). Also, this impact causes the disappearance of many species and otherwise artificially created microhabitats are colonized by other species, perhaps more tolerant to changes or perhaps better adapted to new habitats formed in the river bed and that are not characteristic of their original bed. This may be the case of the Limia River, isolated from the rest and with low faunal affinity, especially in site LI1 (Figure 6), situated upstream the power station, as damming causes the slowdown of the water flow and low levels of oxygen, similar to that observed in stagnant water environments.

In conclusion, in the studied rivers the main factors that determined the macroinvertebrate fauna were (1) water pollution, which directly affected water quality and (2) damming, a disturbance that has caused a change in water flow. In this sense, the negative effect of the anthropogenic impact, especially water contamination, on macroinvertebrates in some of the studied rivers is evident, as shown by the decrease in richness attributes and the IBMWP biological index in impacted sites. As anticipated, there is a loss of species as land use changes from rural to urban. Besides, the water quality in rivers located in natural areas and without strong impacts is better than in rivers located in areas with urban land uses. We also have demonstrated that macroinvertebrates can be used as indicators of environmental impacts in rivers. Their responses to impacts in rivers differ; the majority of taxa are not tolerant to increasing contamination and changes in river structure, but some taxa seem to have adapted to these changes and become dominant in highly disturbed sites. As expected, rare taxa appear to be unmistakeably associated with good water quality, which highlights the importance of conserving freshwater habitats.

Fig. 7. Site LI1 of the Limia River (Ourense, NW Spain), with slow flow caused by damming.

5. References

Abellán, P.; Sánchez-Fernández, D.; Millán, A.; Botella, F.; Sánchez-Zapata, J. A. & Giménez, A. (2006). Irrigation pools as macroinvertebrate habitat in a semi-arid agricultural landscape (SE Spain). *Journal of Arid Environmenats*, 67, 255-269.

Alba-Tercedor, J. & Sánchez-Ortega, A. (1988). Un método rápido y simple para evaluar la calidad biológica de las aguas corrientes basado en el de Hellawell (1978). *Limnetica*, 4, 51–56.

Alba-Tercedor, J. (1996). Macroinvertebrados acuáticos y calidad de las aguas de los ríos. *IV Simposio del Agua de Andalucía (SIAGA)*, 2, 203-213.

Alba-Tercedor, J.; Jáimez-Cuéllar, P.; Álvarez, M.; Avilés, J.; Bonada, N.; Casas, J.; Mellado, A.; Ortega, M.; Pardo, I.; Prat, N.; Rieradevall, M.; Robles, S.; Sáinz-Cantero, C.E.; Sánchez-Ortega, A.; Suárez, M.L.; Toro, M.; Vidalabarca, M.R.; Vivas, S. & Zamora-Muñoz, C. (2002). Caracterización del estado ecológico de ríos mediterráneos ibéricos mediante el índice IBMWP (antes BMWP'). *Limnetica*, 21(3-4), 175-185.

Alonso, A. (2006). Valoración del efecto de la degradación ambiental sobre los macroinvertebrados bentónicos en la cabecera del río Henares. *Ecosistemas*, 15(2), 1-5.

Álvarez-Troncoso, R., (2004). *Estudio faunístico y ecológico de los tricópteros de la cuenca del río Avia (Ourense, España)*, Degree Thesis, Universidad de Vigo, Vigo, Spain.

Argyroudi, A.; Poirazidis, K. & Lazaridou, M. (2010). Macroinvertebrate communities of intermittent and ephemeral streams of Dadia – Lefkimi – Soufli Forest National Park and the impact of small release dams. In: *The Dadia – Lefkimi – Soufli Forest National Park, Greece: Biodiversity, Management and Conservation*, Catsadorakis, G. & Källander, H. (eds). 2010. WWF Greece, Athens.

Armitage, P.D.; Moss, D.; Wright, J.F. & Furse, M.T. (1983). The performance of a new biological water quality score system based on macroinvertebrates over a wide range of unpolluted running water sites. *Water Research*, 17, 333–347.

Artemiadou, V. & Lazaridou, M. (2005). Evaluation Score and Interpretation Index for the ecological quality of running waters in Central and Northern Hellas. *Environmental Monitoring and Assessment*, 10, 1–40.

Artemiadou, V.; Statiri, X.; Brouziotis, T.H. & Lazaridou, M. (2008). Ecological quality of small mountainous Mediterranean streams (river type R-M4) and performance of the European intercalibration metrics. *Hydrobiologia*, 605(1), 75-88.

Baeza, D.; Martínez-Capel, F. & García de Jalón, D. (2003). Variabilidad temporal de caudales: aplicación a la gestión de ríos regulados. *Ingeniería del Agua*, 10(4), 469-478.

Beasley, G. & Kneale, P. (2003). Investigating the influence of heavy metals on macroinvertebrate assemblages using Partial Canonical Correspondence Analysis (pCCA). *Hydrology and Earth Systems Sciences*, 7(2), 221-233.

Beasley, G. & Kneale, P. (2004). Assessment of heavy metal and PAH contamination of urban streambed sediments on macroinvertebrates. *Water, Air and Soil Pollution*, 4, 563-578.

Belmar, O.; Velasco, J.; Martínez-Capel, F. & Marín, A.A. (2010). Natural flow regime, degree of alteration and environmental flows in the Mula stream (Segura River basin, SE Spain). *Limnetica*, 29(2), 353-368.

Benejam, L.; Angermeier, P.L.; Munné, A. & García-Berthou, E. (2010). Assessing effects of water abstraction on fish assemblages in Mediterranean streams. *Freshwater Biology*, 55: 628-642.

Benetti, C.J. & Garrido, J. (2010). The influence of stream habitat and water quality on water beetles assemblages in two rivers in northwest Spain. *Vie et milieu*, 60(1), 53-63.

Blasco, J.; Sáenz, V. & Gómez-Parra, A. (2000). Heavy metal fluxes at the sediment-water interface of three coastal ecosystems from south-west of the Iberian Peninsula. *Science of the Total Environment*, 247, 189-199.

Bonada, N.; Dallas, H.; Rieradevall, M.; Prat, N. & Day, J. (2006a). A comparison of rapid bioassessment protocols used in 2 regions with Mediterranean climates, the Iberian Peninsula and South Africa. *Journal of the North American Benthological Society*, 25(2), 487-500.

Bonada, N.; Prat, N.; Resh, V.H. & Statzner, B. (2006b). Devolopments in aquatic insect biomonitoring: A Comparative Analysis of Recent Approaches. *Annual Reviews of Entomology*, 51, 495 – 523.

Bonada, N.; Rieradevall, M. & Prat, N. (2000). Temporalidad y contaminación como claves para interpretar la biodiversidad de macroinvertebrados en un arroyo mediterráneo (Riera de Sant Cugat, Barcelona). *Limnetica*, 18, 81-90.

Bredenhand, E. & Samways, M.J. (2009). Impact of a dam on benthic macroinvertebrates in a small river in a biodiversity hotspot: Cape Floritic Region, South Africa. *Journal of Insect Conservation*, 13(3), 297-307.

Brinkhurst, R.O. & Gelder, S.R. (2001). Annelida: Oligochaeta inlcuding Branchiobdellidae, In: *Ecology and classification of North American freshwater invertebrates*, Thorp, J. P. & Covich, M. P., 431-464, Academic Press, Orlando, USA.

Buffagni, A. & Furse, M. (2006). Intercalibration and comparison – major results and conclusions from the STAR project. *Hydrobiologia*, 566, 357–364.

Chatzinikolaou, Y. & Lazaridou, M. (2007). Identification of the self-purification stretches of the Pinios River, Central Greece. *Mediterranean Marine Science*, 8 (2), 19-32.

Cheimonopoulou M.; Bobori D.; Theocharopoulos I. & Lazaridou M. (2011). Assessing Ecological Water Quality with Macroinvertebrates and Fish: A CaseStudy from a Small Mediterranean River. *Environmental Management*, 47, 279–290.

Chessman, B.C.; Growns, J.E. & Kotlash, A.R. (1997). Objective derivation of macroinvertebrate family sensitivity grade numbers for the SIGNAL biotic index: application to the Hunter River system, New South Wales. *Australian Journal of Marine and Freshwater Research*, 48, 159-172.

Council of the European Communities (2000). Directive 2000/60/EC of the European Parliament and of the Council of 23 October 2000 establishing a framework for Community action in the field of water policy. *Official Journal of the European Communities*, L327, 1-72.

Dahl, J.; Johnson, R.K. & Sandin, L. (2004). Detection of organic pollution of streams in southern Sweden using benthic macroinvertebrates. *Hydrobiologia*, 516, 161-172.

Dallas, H.F. (1995). *An evaluation of SASS (South African Scoring System) as a tool for the rapid bioassessment of water quality*, MSc Thesis, University of Cape Town, Cape Town, South Africa.

Dallas, H.F. (1997). A preliminary evaluation of aspects of SASS (South African Scoring System) for the rapid bioassessment of water quality in rivers, with particular reference to the incorporation of SASS in a national biomonitoring programme. *South African Journal of Aquatic Sciences*, 23, 79–94.

De Pauw, N. & Vanhooren, G. (1983). Method for biological quality assessment of watercourses in Belgium. *Hydrobiologia*, 100, 153-168.

Dessaix, J.; Fruget, J.F.; Olivier, J.M. & Beffy, J.L. (1995). Changes of the macroinvertebrate communities in the dammed and by-passed sections of the French upper Rhône after regulation. *Regulated Rivers: Research & Management*, 10, 265-279.

Dolédec, S.; Dessaix, J. & Tachet, H. (1996). Changes within the Upper Rhône River macrobenthic communities after the completion of three hydroelectric schemes: anthropogenic effects or natural change? *Archiev für Hydrobiologie*, 136, 19-40.

Dudgeon, D.; Arthington, A.H.; Gessner, M.O.; Kawabata, Z.I.; Knowler, D.J.; Lévêque, C.; Naiman, R.J.; Prieur-Richard, A.H.; Soto, D.; Stiassny, M.L.J. & Sullivan, C.A. (2006). Freshwater biodiversity: importance, threats, status and conservation challenges. *Biological Reviews of the Cambridge Philosophical Society*, 81, 163-182.

Elbaz-Poulichet, F.; Morley, N.H.; Cruzado, A.; Velasquez, Z.; Achterberg, E.P. & Braungardt, C.B. (1999). Trace metal and nutrient distribution in an extremely low pH (2.5) river-estuarine system, the Ria of Huelva (South-West Spain). *Science of the Total Environment*, 227, 73-83.

EN 27828 (1994). Water quality - Methods of biological sampling - *Guidance on hundnet sampling of aquatic benthic macro-invertebrates* (ISO 7828:1985).

European Commission, (2008). Commission decision of 30 October 2008 establishing, pursuant to Directive 2000/60/EC of the European Parliament and of the Council, the values of the Member State monitoring system classifications as a result of the intercalibration exercise (notified under document number C (2008) 6016) (Text with EEA relevance) (2008/915/EC). *Official Journal of the European Communities* L332/20, Luxemburg.

European Commission, (2010). Common Implementation Strategy for the Water Framework Directive (2000/60/EC). Guidance document no. 14. Guidance on the intercalibration process 2008 – 2011. http://circa.europa.eu/Public/irc/env/wfd/library?l=/framework_directive/gui dance_documents/intercalibration_1/_EN_1.0_&a=d

Fernández Díaz, M. (2003). *Estudio faunístico y ecológico de los coleópteros acuáticos (Adephaga y Polyphaga) en la cuenca del río Ávia (Ourense, NO España): Distribución espacial y temporal*, Degree Thesis, Universidad de Vigo, Vigo, Spain.

Fernández-Díaz, M.; Benetti, C.J. & Garrido, J. (2008). Influence of iron and nitrate concentration in water on aquatic Coleoptera community structure: application to the Avia river (Ourense, NW, Spain). *Limnetica*, 27 (2), 285-298.

Fjellheim, A. & Raddum, G.G. (1996). Weir building in a regulated west Norwegian river: longterm dynamics of invertebrates and fish. *Regulated Rivers: Research & Management*, 12, 501-508.

Fjellheim, A.; Havardstun, J.; Radum, G.G. & Schenell, O.A. (1993). Effects of increased discharge on benthic macroinvertebrates in a regulated river. *Regulated Rivers: Research & Management*, 8, 179-187.

Freund, G.J. & Petty, J.T. (2007). Response of fish and macroinvertebrate bioassessment indices to water chemistry in a mined Appalachian Watershed. *Environmental Management*, 39, 707-720.

Fulan, J.A.; Raimundo, R.; Figueiredo, D. & Correia, M. (2010). Abundance and diversity of dragonflies four years after the construction of a reservoir. *Limnetica*, 29(2), 279-286.

Gabriels, W.; Goethals, P.L.M. & De Pauw, N. (2005). Implications of taxonomic modifications and alien species on biological water quality assessment as exemplified by the Belgian Biotic Index method, *Hydrobiologia*, 542, 137–150.

García-Criado, F. Fernández Aláez, M. (2001). Hydraenidae and Elmidae assemblages (Coleoptera) from Spanish river basin: good indicators of coal mining pollution? *Archiv für Hydrobiologie* 150, 641-660.

García-Criado, F.; Fernández Aláez, C. & Fernández Aláez, M. (1999). Environmental variables influencing the distribution of Hydraenidae and Elmidae assemblages (Coleoptera) in a moderately-polluted river basin in north-western Spain. *European Journal of Entomology*, 96, 37-44.

Garrido, J.; Alonso, A.I.; Álvarez-Troncoso, R.; Benetti, C.; Fernández-Díaz, M. & Gayoso, A. (2003). *Programa de vigilancia ambiental de la minicentral del Río Deva (Pontevedra)*, Technical Report, Universidad de Vigo, Vigo, Spain.

Garrido, J.; Alonso, A.I.; Álvarez-Troncoso, R.; Benetti, C.; Fernández-Díaz, M. & Gayoso, A.; González, A.; Penín, L.; Pérez-Bilbao, A.; Rodríguez, I. & Vidal, M. (2005). *Programa de vigilancia ambiental de la minicentral de Panteliñares (Ourense)*, Technical Report, Universidad de Vigo, Vigo, Spain.

Garrido, J.; Alonso, A.I.; Álvarez-Troncoso, R.; Fernández-Díaz, M. & Vidal, M. (2000a). *Evaluación de la calidad del agua mediante parámetros biológicos y fisicoquímicos en los ríos de la cuenca del Avia (Ourense)*, Technical Report, Universidad de Vigo, Vigo, Spain.

Garrido, J.; Alonso, A.I.; Álvarez-Troncoso, R.; Fernández-Díaz, M.; Vidal, M. & Benetti, C. (2002). *Estudio y seguimiento de la recuperación biológica de la cuenca del río Lagares (Vigo, Pontevedra)*, Technical Report, Universidad de Vigo, Vigo, Spain.

Garrido, J.; Benetti, C.J. & Pérez-Bilbao, A. (2008). *Evaluación de la calidad del agua del Río Furnia (Tuy, pontevedra) mediante indicadores biológicos, antes de la puesta en marcha de la estación depuradora de aguas residuales*, Technical Report, Universidad de Vigo, Vigo, Spain.

Garrido, J.; Membiela, P. & Vidal, M. (1998). Calidad biológica de las aguas del río Barbaña. *Tecnología del agua*, 175, 50-54.

Garrido, J.; Vidal, M.; Alonso, A.I.; Cuadrado, D.; Outerelo, J.A. & Fernández-Díaz, M. (1999). *Programa de vigilancia ambiental del aprovechamiento hidroeléctrico de fecha en el Río Tambre (A Coruña). Fase de funcionamiento*, Technical Report, Universidad de Vigo, Vigo, Spain.

Garrido, J.; Vidal, M.; Alonso, A.I.; Cuadrado, D.; Outerelo, J.A. & Fernández-Díaz, M. (2000b). *Programa de vigilancia ambiental del aprovechamiento integral del Río Tuño (Ourense)*, Technical Report, Universidad de Vigo, Vigo, Spain.

Garrido, J.; Vidal, M.; Alonso, A.I.; Cuadrado, D.; Outerelo, J.A. & Fernández-Díaz, M. (2000c). *Programa de vigilancia ambiental de la minicentral hidroeléctrica del Río Deva (Ourense)*, Technical Report, Universidad de Vigo, Vigo, Spain.

Garrido, J.; Vidal, M.; Alonso, A.I.; Gayoso, A.; Cuadrado, D.; Outerelo, J.A. & Fernández-Díaz, M. (2000d). *Programa de vigilancia ambiental de la minicentral del Río Tea, propiedad de Engasa,* Technical Report, Universidad de Vigo, Vigo, Spain.

Gerhardt, A. (2000). *Biomonitoring of Polluted Water: Reviews on Actual Topics.* Trans Tech Publications Inc., Zürich, Switzerland.

Harper, M.P. & Peckarsky, B.L. (2005). Effects of pulsed and pressed disturbances on the benthic invertebrate community following a coal spill in a small stream in northeastern USA. *Hydrobiologia,* 544, 241-247.

Hauer, F.R. & Resh., V.H. (1996). Benthic Macroinvertebrates, In: *Methods in Stream Ecology,* F.R. Hauer & G.A. Lamberti (eds), pp. 339-369, Academy Press, New York, USA.

Heino, J. (2000). Lentic macroinvertebrate assemblage structure along gradients in spatial heterogeneity, habitat size and water chemistry. *Hydrobiologia,* 418, 229-242.

Hirst, H.; Jüttner, I. & Ormerod, S.J. (2002). Comparing the responses of diatoms and macroinvertebrates to metals in upland streams of Wales and Cornwall. *Freshwater Biology,* 47, 1752-1765.

Jenkins, A.; McCartney, M. & Sefton, C. (1993). Impacts of climate change on river quality in the United Kingdom. *Report to Department of Environment. Institute of Hydrology.* Wallingford.

Jesus, T.; Formigo, N.; Santos, P. & Tavares, G.R. (2004). Impact evaluation of the Vila Viçosa small hydroelectric Power plant (Portugal) on the water quality and on the dynamics of the benthic macroinvertebrate communities of the Ardena river. *Limnetica,* 23 (3-4), 241-256.

Johnson, R.K.; Wiederholm, T. & Rosenberg, D.M. (1993). Freshwater biomonitoring using individual organisms, populations, and species assemblages of benthic macroinvertebrates, In: *Freshwater Biomonitoring and Benthic Macroinvertebrates,* D.M. Rosenberg & V.H. Resh (eds), pp. 40-158, Chapman & Hall, New York, USA.

Kottek, M.; Grieser, J.; Beck, C.; Rudolf, B. & Rubel, F. (2006). World Map of the Köppen-Geiger climate classification updated. *Meteorologiche Zeitung,* 15, 259-263.

Kubecka, J.; Matena, J. & Hartvich, P. (1997). Adverse ecological effects of small hydropower stations in the Czech Republic: 1. Bypass plants. *Regulated rivers: research & management,* 13, 1001-113.

Lessard, J.L. & Hayes, D.B. (2003). Effects of elevated water temperature on fish and macroinvertebrate communities below small dams. *River Research Applications,* 19, 721-732.

Lytle, D.A. & Peckarsky, B.L. (2001). Spatial and temporal impacts of a diesel fuel spill on stream invertebrates. *Freshwater Biology,* 46, 693-704.

Marqués, M.J.; Martínez-Conde, E. & Rovira, J.V. (2003). Effects of zinc and lead mining on the benthic macroinvertebrates of a fluvial ecosystem. *Water, Air, and Soil Pollution,* 148, 363 - 388.

McCabe, D.J. & Gotelli, N.J. (2000). Effects of disturbance frequency, intensity, and area on assemblages of stream macroinvertebrates. *Oecologia,* 124, 270-279.

Metcalfe, J.L. (1989). Biological quality assessment of running waters based on macroinvertebrate communities: History and Present Status in Europe. *Environmental Pollution,* 60, 101-139.

Ministry of Environment (2005). *Metodología para el establecimiento del estado ecológico según la Directiva Marco del Agua. Protocolos de muestreo y análisis para invertebrados bentónicos,* Madrid, España.

National Water Council (1981). *River quality: the 1980 survey and future outlook,* London, UK.

Nummelin, M.; Lodenius, M.; Tulisalo, E.; Hirvonen, H. & Alanki, T. (2007). Predatory insects as bioindicators of heavy metal pollution. *Environmental Pollution,* 145, 339-347.

Oller, C. & Goitia, E. (2005). Benthic macroinvertebrate and heavy metals in the Pilcomayo River (Tarija, Bolivia). *Revista Boliviana de Ecología,* 18, 17-32.

Ortiz, J.D.; Martí, E. & Puig, M.A. (2005). Recovery of the macroinvertebrate community below a wastewater treatment plant input in a Mediterranean stream. *Hydrobiologia,* 545, 289–302.

Oscoz, J.; Campos, F. & Escala, M.C. (2006). Variación de la comunidad de macroinvertebrados bentónicos en relación con la calidad de las aguas. *Limnetica,* 25 (3), 683-692.

Paz, C. (1993). *Hydradephaga (Coleoptera) en la cuenca del río Landro (NW Península Ibérica). Estudio faunístico y ecológico,* Ph. D. Thesis, Universidad de Santiago de Compostela, Santiago de Compostela, Spain.

Pérez-Bilbao, A. & Garrido, J. (2009). Evaluación del estado de conservación de una zona LIC (Gándaras de Budiño, Red Natura, 2000) usando los coleópteros acuáticos como indicadores. *Limnetica,* 28 (1), 11-22.

Pérez-Quintero, J.C. (2007). Diversity, habitat use and conservation of freshwater molluscs in the lower Guadiana River basin (SW Iberian Peninsula). *Aquatic Conservation: Marine and Freshwater Ecosystems,* 17, 485–501.

Poquet, J.M.; Alba-Tercedor, J.; Puntí, T.; Sánchez-Montoya, M.M.; Robles, S.; Álvarez, M.; Zamora-Muñoz, C.; Sáinz-Cantero, C.E.; Vidal-Abarca, M.R.; Suárez, M.L.; Toro, M.; Pujante, A.M.; Rieradevall, M. & Prat, N. (2009). The MEDiterranean Prediction And Classification System (MEDPACS): an implementation of the RIVPACS/ AUSRIVAS predictive approach for assessing Mediterranean aquatic macroinvertebrate communities. *Hydrobiologia,* 623, 153–171.

Prat, N.; Munné, A.; Solá, C.; Rieradevall, M.; Bonada, N. & Chacón, G. (1999). *La qualitat ecològica del Llobregat el Besòs i el Foix,* Àrea de Medi Ambient, Diputació de Barcelona, Barcelona, España.

Prenda, J. & Gallardo-Mayenco, A. (1996). Self-purification, temporal variability and the macroinvertebrates community in small lowland Mediterranean streams receiving crude domestic sewage effluents. *Archiv für Hydrobiologie,* 136, 159-170.

Quinn, J.M.; Cooper, A.B.; Davies-Colley, R.J.; Rutherdford, J.C. & Williamson, R.B. (1997). Land use effects on habitat, water quality, periphyton, and benthic invertebrates in Waikato, New Zealand, hill-country streams. *New Zealand Journal of Marine and Freshwater Research,* 31, 579-597.

Rahel, F.J. (2002). Homogenization of freshwater faunas. *Annual Review of Ecology and Systematics,* 33, 291–315.

Richter, B.D.; Braun, D.P.; Mendelson, M.A. & Master, L.L. (1997). Threats to imperilled freshwater fauna. *Conservation Biology,* 11(5), 1081-1093.

Macroinvertebrates as Indicators of Water Quality in Running Waters: 10 Years of Research in Rivers with Different
Degrees of Anthropogenic Impacts

71

Rosenberg, D.M. & Resh, V.H. (1993). Introduction to freshwater biomonitoring and benthic macroinvertebrates, In: *Freshwater Biomonitoring and Benthic Macroinvertebrates*, D.M. Rosenberg & V.H. Resh (eds), pp. 1-9, Chapmann & Hall, New York, USA.

Sarr, A.B. (2011). *Evaluación del estado de conservación de ríos afectados por minicentrales hidroeléctricas mediante el estudio de coleópteros acuáticos*. PhD Thesis, University of Vigo, Vigo, Spain.

Smolders, A.J.P.; Lock, R.A.C.; Van der Velde, G.; Medina Hoyos, R.I. & Roelofs, J.G.M. (2003). Effects of mining activities on heavy metal concentrations in water, sediment, and macroinvertebrates in different reaches of the Pilcomayo River, South America. *Archives of Environmental Contamination and Toxicology*, 44, 314-323.

Smolders, A.J.P.; Van Hengstum, G.; Loermans, J.; Montes, A. Rizo, H. &. Castillo, I. (1999). Efectos de la Contaminación Minera sobre la Composición de la Macrofauna Bentónica en el Río Pilcomayo. *Revista Boliviana de Ecología*, 6,229-237.

Song, M.Y.; Leprieur, F.; Thomas, A.; Lek-Ang, S.; Chon, T.S. & Lek, S. (2009). Impact of agricultural land use on aquatic insect assemblages in the Garonne river catchment (SW France). *Aquatic Ecology*, 43, 999-1009.

Stanford, J.A. & Ward, J.V. (1979). Stream regulation in North America, In: *The ecology of regulated streams*, J.V. Ward & J.A. Stanford (eds.), pp. 215-236, Plenum Press, New York, USA.

Stanley, E.H.; Luebke, M.A.; Doyle, M.W. & Marshll, D.W. (2002). Short-term changes in channel form macroinvertebrate communities following low-head dam removal. *Journal of the North American Benthological Society*, 21, 172-187.

Strayer, D.L. (2006). Challenges for freshwater invertebrate conservation. *Journal of the North American Benthological Society*, 25, 271-287.

Ter Braak, C.J.F. & Šmilauer, P. (2002). *CANOCO: Referente manual and CanoDraw for Windows User's guide: Software for Canonical Community Ordination (version 4.5)*. Microcomputer Power, Ithaca.

Thomson, J.R.; Hart, D.D.; Charles, D.F.; Nightengale, T.L. & Winter, D.M. (2005). Effects of removal of a small dam on downstream macroinvertebrate and algal assemblages in a Pennsylvania stream. *Journal of the North American Benthological Society*, 24(1), 192-207.

Thorne, R.J. & Williams, W.P. (1997). The response of benthic macroinvertebrates to pollution in developing countries: a multimetric system of bioassessment. *Freshwater Biology*, 37, 671-686.

Tonkin, J.D.; Death, R.G. & Joy, M.K. (2009). Invertebrate drift patterns in a regulated river: dams, periphyton biomass or longitudinal patterns? *River Research and Applications*, 25(10), 1205-1338.

Velasco, J.; Millán, A.; Hernández, J.; Gutiérrez, C.; Abellán, P.; Sánchez, D. & Ruiz, M. (2006). Response of biotic communities to salinity changes in a Mediterranean hypersaline stream. *Saline Systems*, 2 (12), 1-15.

Woodiwis, F.S. (1964). The biological system of stream classification used by Trent River Board. *Chemistry & Industry*, 11, 443-447.

Yoshimura, M. (2008). Longitudinal patterns of benthic invertebrates along a stream in the temperate forest in Japan: in relation to humans and tributaries. *Insect Conservation and Diversity*, 1, 95–107.

Zamora-Muñoz, C. & Alba-Tercedor, J. (1996). Bioassessment of organically polluted Spanish river, using a biotic index and multivariate methods. *Journal of the North American Benthological Society,* 15 (3), 332- 352.

4

Biofilms Impact on Drinking Water Quality

Anca Farkaş[1], Dorin Ciataraş[1] and Brânduşa Bocoş[2]
[1]Someş Water Company,
[2]National Public Health Institute, Regional Public Health Center of Cluj,
Cluj-Napoca,
Romania

1. Introduction

A paradigm shift currently occurs in microbiology, with significant impacts in a variety of environmental, medical and industrial applications. The old misconception of free floating microbes is invalidated by a different knowledge pattern: the great majority of terrestrial microorganisms live in communities associated to surfaces, called biofilms (Costerton et al, 1987; Flemming, 2008; Muntean, 2009). This organisation mode is associated to all surfaces in contact with water in drinking water processing, storage and distribution. Such biofilms are represented by structured consortia of sessile microorganisms characterized by surface attachment, self-produced exopolymeric matrix, structural, functional and metabolic heterogeneity, capable of intercellular communication by quorum-sensing and plurispecific composition.

Biofouling in drinking and industrial water systems has detrimental effects such as microbiological and chemical deterioration in water quality, corrosion inducing, drinking water treatment yield loss, efficiency reducing in cooling and heating exchange and transport, as well as in membrane processes (White et al, 1999; Flemming et al, 2002; LeChevallier and Au, 2004; Coetser and Cloete, 2005).

Biofilms are playing a major role in drinking and waste water treatment processes due to their enhanced properties of mineralization, bioaccumulation and bioadsorbtion. Despite the beneficial effects of the biological filter known as *schmutzdecke* in slow sand filtration or of the bio-sand filters, biofilms occurrence in other treatment stages, in drinking water networks and reservoirs represents a continuous challenge to water professionals. Drinking water associated biofilms induce residual disinfectants depletion and may cause aesthetic problems consisting in colour, odour and taste degradation due to chemical compounds released and more important, they pose a threat to human and animal health by hosting pathogenic or toxins producing bacteria, viruses, protozoa, algae, fungi and invertebrates. The great majority of water related health problems are the result of microbial contamination (Riley et al, 2011). Considering these aspects, naturally occurring biofilms in contact with drinking water were identified and described as microbial reservoirs for further contamination (Szewzyk et al, 2000; Wingender and Flemming, 2011).

The complex structure of drinking water associated biofilms is influenced by the microbial composition of source water and sediments (LeChevallier et al, 1987; Szewzyk et al, 2000;

Emtiazi et al, 2004). They may enter the distribution network, escaping the treatment and disinfection processes (known as breakthrough) and multiply in bulk water or biofilms. The two modes of multiplication are defined as regrowth (recovery of disinfectant injured cells) and aftergrowth (microbial growth in a distribution system) processes (Characklis, 1988; van der Kooij, 2003). Pathogenic microorganisms of concern may also emerge in drinking water systems by intrusion, due to external contamination events in different steps of water treatment, storage and transportation: cross connections, backflow events, pipe breaks, negative pressure and because of improper flushing and disinfection procedures.

2. Drinking water biofilms, emerging pathogens and opportunistic pathogens

The most alarming consequences as a result of biofouling in drinking water distribution systems consist in the presence, multiplication and dispersion into water of bacterial pathogens, opportunistic pathogens, parasitic protozoa, viruses and toxins releasing fungi and algae. They may appear as primary colonizers promoting the adhesion at the interface and subsequent biofilm formation (Costerton, 1994), but more often as secondary colonizers in ecological microniches offered by the existent attached community.

Emerging pathogens are those that have appeared in a human population for the first time, or have occurred previously but are increasing in incidence or expanding to areas where they have not previously been reported, usually over the last 20 years. They include: bacteria (pathogenic *E. coli, Helicobacter pylori, Campylobacter jejuni, Mycobacterium* avium complex), parasitic protozoa (*Cryptosporidium* spp., *Cyclospora cayetanensis, Toxoplasma gonidii*), viruses (noroviruses, hepatitis E) and toxic cyanobacteria (Hunter et al, 2003). Opportunistic pathogens are commonly members of water microbiota that would be normally harmless to a healthy individual but can infect a compromised host (US EPA, 2002).

Such microorganisms were detected worldwide in drinking water and associated biofilms and in raw water and sediments. In this context, establishing health-based targets, drinking water quality assurance implies effective preventive measures, such as sources water protection in order to reduce contamination risk in these strategic environments, as well as corrective actions stipulated in water safety plans prepared by water suppliers.

When cases of illness are registered, epidemiological studies are conducted in order to demonstrate similarities in genetic profiles of strains isolated from clinical and the environmental specimens, to track the source of infection. Drinking water and associated biofilms are often among the prime candidates tested when gastrointestinal diseases and different types of infections are recorded.

Developing countries are facing a major lack of safe drinking water, the transmission via faecal-oral route causing enormous numbers of severe water related illness and cases of deaths, especially in infants (Riley et al, 2011). Microbial source tracking approach is particularly important for drinking water sources, in order to identify the origin of contamination. Increasing population's access to clean drinking water and sanitation facilities is one of the first priorities of local and global authorities.

Even if the access to high quality drinking water is provided, neither humans in rich countries have definitely won the fight against microbes, yet. Tap water quality assurance is facing new challenges, consisting mainly in biofouling issues, emergent waterborne pathogens, toxins releasing and opportunistic pathogens occurrence. Besides households, different branches of industry, especially food and pharmaceutical, health care facilities, schools, nursing homes and other critical areas are carefully included in monitoring assays. Recently is seriously considered the ability of some species, referred as opportunistic pathogens, to induce disease under certain circumstances in immunocompromised individuals: immunosuppressed, malnourished, diabetic, burn, cancer, AIDS, on haemodialysis, respiratory, with organ transplants patients (Rusin et al, 1997; Payment and Robertson, 2004). Sensitive subpopulations such as young children, elderly persons or pregnant women are also vulnerable to infections caused by opportunistic pathogens (Reynolds et al, 2008). Other special categories of exposed subjects consist in patients with indwelling cannulae and catheters, implant devices and contact lenses wearers. Opportunistic pathogens are becoming a major issue, causing from allergy or superficial infections to life-threatening systemic infections, since the ascending trend in congenital and acquired immunodeficiency affecting global population. Such species presence is often investigated, in addition to routine monitoring, within drinking water and associated biofilms, as wide-occurring bacteria of concern in the continuous increasing category of hospitalised and ambulatory immunocompromised persons (Glasmacher et al, 2003).

Even in drinking water carefully treated and distributed at high standards, pathogenic contamination and disease outbreaks may occur (Szewzyk et al, 2000; Wingender & Flemming, 2011) demonstrating the imperative requirement for comprehensive water safety plans implementation.

2.1 Drinking water quality assessment – Microbiological aspects and biofilms

Tap water supposes not to be and is not sterile, microbial load in bulk water consisting mainly in inoffensive heterotrophs, presumably coming from associated biofilms by detachment during dispersion. Routine monitoring of raw water sources, finishing water at the exit from treatment plant, drinking water in pipe networks, service reservoirs and finally at the consumers implies periodically investigations of a number of water samples collected with a frequency depending on the population deserved. According to European regulations, microbial indicators assessed by standardised conventional culturing techniques are: colony count at 37°C, colony count at 22°C, total coliforms, *Escherichia coli*, intestinal enterococci and *Clostridium perfringens*. The greatest microbial risk being associated with ingestion of water contaminated with human or animal faeces, thus the potential presence of pathogenic bacteria, viruses and cysts of protozoan parasites; faecal indices (*E. coli*, intestinal enterococci and *C. perfringens*) presence is routinely investigated.

The shortcomings of water quality monitoring based on faecal indicators and heterotrophic plate count, resulting in underestimation of drinking water microbial populations in numbers and composition are discussed worldwide considering the following:

- Only a small volume approximated to represent from 2×10^{-7} to $5 \times 10^{-7}\%$ of delivered drinking water is examined in routine monitoring (Allen, 2011);
- In drinking water systems, the high majority of bacteria, estimated at 95%, are located attached at the surfaces, while only 5% are found in water phase and detected by sampling as commonly used for quality control (Flemming et al, 2002). Other studies

are indicating bacterial numbers characterizing the biomass in pipe biofilms being 25 times more abundant than the suspended cells (Servais et al, 2004). Although, the common notion that biofilms dominate the distribution systems has been proven to be not true under all conditions by Srinivasan (2008), whose findings suggest that bulk bacteria may dominate in network sections containing chlorine residuals lower than 0.1mg/L and having residence time longer than 12 hours.

- A significant percent of water and biofilm bioburden may be in a viable but non cultivable state, unable to grow on artificial growth media but alive and capable of renewed activity and so hygienically relevant (Moritz et al, 2010). A small fraction of waterborne microorganisms (0.01%) are estimated to be culturable heterotrophic bacteria (Watkins and Jian, 1997; Exner et al, 2003);

- Limitations of detection methods (Lehtola et al, 2006; 2007; September et al, 2007). The investigation of drinking water associated biofilms from four European countries (France, Great Britain, Portugal and Latvia) confirmed *E. coli* presence by culturing techniques in one out of five pipes whereas all networks except one were positive for *E. coli* using the PNA FISH methods; their viability was also demonstrated. *E.coli* contributed with percents from 0.001% to 0.1% in the total bacterial numbers (Juhna et al., 2007);

- Faecal indicators are the best predictors of potential risks, but their concentrations rarely correlate perfectly with those of pathogens (Payment and Locas, 2011). Although in freshwater significant correlations have been established between faecal indices and pathogenic species, their presence in drinking water showed limited or no correlation with different species of pathogenic or opportunistic pathogenic bacteria, viruses, protozoa and fungi. Water quality assessment based only on the investigation of faecal indicators' presence proved to be insufficient when many waterborne outbreaks emerged. Still, until more reliable indices and methods of detection will be wider implemented, the well- known standardised procedures are applied in routine monitoring across the globe.

Many studies targeting attached microbial communities have been performed for quantification of the total number of germs, by different methods. They offer an unspecific overview upon microbial load, bringing certain information about drinking water treatment process efficiency and distribution system integrity. Still, the real composition and dynamics of microbial populations within drinking water associated biofilms represents a continuous challenge. Experimental biofilm succession monitored for a long term development indicated a stable population state after 500 days in a model drinking water distribution system. A homogenous composition of the population in the mature biofilm could mask a dynamic situation at a smaller scale (Martiny et al., 2003). Quantitative and prescriptive evaluation is the next target of scientific community. Prediction of microorganisms' behaviour in the distribution system water and biofilms requires greater understanding of the effects in microbial attachment, detachment, survival, multiplication and viability of three groups of abiotic and biotic factors: substratum physicochemical properties (type of materials), biofilm composition (microbial intra- and interspecific interactions) and bulk water characteristics (disinfectants residuals, oxygen and nutrients concentrations, system hydraulics, temperature).

2.1.1 Occurrence of bacteria in drinking water and associated biofilms

Among the nuisance bacteria regularly found in drinking water and biofilms, species that are not characteristic to the water environment may appear due to contamination events, with major impacts upon human health. Enteric bacteria such as *Escherichia coli, Klebsiella pneumoniae, K. oxytoca, Enterobacter cloacae, E. aglomerans, Helicobacter pylori, Campylobacter* spp., *Shigella* spp., *Salmonella* spp., *Clostridium perfringens, Enterococcus faecalis, E. faecium,* as well as environmental bacteria becoming opportunistic pathogens *Legionella pneumophila, Pseudomonas aeruginosa, P. fluorescens, Aeromonas hydrophila, A. caviae, Mycobacterium avium, M. xenopi,* together with other waterborne agents have been indicated to live in ecological microniches offered by drinking water associated biofilms (table 1).

When compared with planktonic counterparts, biofilm bacteria and other inhabitants display superior characteristics due to specialization within this emergent structure and to complex relationships established (Costerton, 1994). Community belonging, from a microbial perspective represents a benefit materialized in increasing the chances of survival in this oligotrophic environment by offering ecological microniches, establishing intra- and interspecific cooperation relationships by communication via quorum sensing and perpetuating individuals' resistance to disinfection agents. Even species not able to survive and most of them incapable of growth and multiplication in water were identified in associated biofilms; recent studies have demonstrated their ability to grow in those microniches. For example, *Legionella pneumophila* survives but does not multiply in sterile drinking water, its proliferation being dependent on parasitic relationship with other microorganisms: 14 species of amoebae, two species of ciliated protozoa, and one slime mould - *L. pneumophila* being described as protozoonotic bacteria (Murga et al, 2001; Fields et al, 2002; Declerck, 2010). In many outbreaks, the presence of pathogenic bacteria was not detected by routine monitoring, the correlation with faecal indicators found in tap water samples being defective. *E. coli* bacillus, the most popular faecal indicator, was chosen inter alia based on its incapacity of growth in water. Recent studies have shown its ability to multiplicate in drinking water associated biofilms under strictly anaerobic conditions (Latimer et al., 2010), so the indicative value of the faecal index of choice becomes questionable.

One of the advantages offered by drinking water biofilm organization to its members is represented by the enhanced resistance to disinfection residuals. The four hypothetical mechanisms of biofilm resistance involve slow antimicrobial penetration, deployment of adaptative stress responses, physiological heterogeneity in biofilm population and the presence of phenotypic variants or persister cells (Chambless et al, 2005). Another benefit from the microbial perspective consists in the emergence of genetically encoded resistance to biocides and antibiotics, and the spread of antimicrobial resistance genes in bacterial populations via mobile genetic elements, by lateral gene transfer. Integrons are genetic elements possessing a site-specific recombination system for assembling of resistance genes in gene cassettes. They play a major role in the rapid spread of antibiotic resistance in clinical environments. Gene cassettes encoding resistance to quaternary ammonium compounds (*qac*) and integron-integrase (*intI*) genes characteristics for class 1 integron were recently recovered from environmental samples, including biofilms from a groundwater treatment plant (Gillings et al, 2009). The proximity of individuals in biofilm consortia and the extremely short generation times in bacteria multiplication are prerequisites for intensive rates of lateral gene transfer and thus resistance spreading and perpetuation in diverse natural or artificial ecosystems.

Bacteria	Samples type/Origin	Country	References
Escherichia coli	Biofilm - WDS	USA	LeChevallier et al, 1987
	Biofilm - WDS	Germany	Schmeisser et al, 2003
	Biofilm – WDS	France, England, Portugal, Latvia	Juhna et al, 2007
	Biofilm, Water - WDS	South Africa	September et al, 2007
	Biofilm - DWTP	Romania	Farkas et al, 2011
Faecal enterococci	Water - WDS	Korea	Lee et al, 2006
	Water - WDS	South Africa	September et al, 2007
	Biofilm - WDS	Portugal	Menaia et al, 2008
	Biofilm - DWTP	Romania	Farkas et al, 2011
Clostridium spp.	Biofilm - WDS	Portugal	Menaia et al, 2008
	Water - WDS	Greece	Kormas et al, 2010
	Biofilm - DTP	Romania	Farkas et al, 2011
Klebsiella spp.	Biofilm - WDS	USA	LeChevallier et al, 1987
	Water - WDS	Korea	Lee et al, 2006
	Biofilm, Water - WDS	South Africa	September et al, 2007
Pseudomonas spp.	Biofilm - WDS	Germany	Schmeisser et al, 2003
	Biofilm - WDS	Germany	Emtiazi et al, 2004
	Water - WDS	Korea	Lee et al, 2006
	Biofilm, Water - WDS	South Africa	September et al, 2007
	Biofilm - WDS	Portugal	Menaia et al, 2008
	Biofilm - DWTP	Romania	Farkas et al, 2011
Aeromonas spp.	Water - WDS	Scotland	Gavriel et al, 1998
	Biofilm, Water - WDS	USA	Chauret et al, 2001
	Biofilm - WDS	Australia	Bomo et al, 2004
	Water - WDS	Korea	Lee et al, 2006
	Biofilm, Water - WDS	South Africa	September et al, 2007
	Water - WDS	Brasil	Razzolini et al, 2008
	Biofilm - DWTP	Romania	Farkas et al, 2011
Vibrio spp. *V. cholerae*	Water - WDS	Korea	Lee et al, 2006
	Biofilm, Water - WDS	South Africa	September et al, 2004; 2007
	Biofilm - pond	Bangladesh	Alam et al, 2007
	Water - reservoirs	Sudan	Shanan et al, 2011
Mycobacterium spp.	Biofilm - WDS	Germany	Schmeisser et al, 2003
	Biofilm - WDS	South Africa	September et al, 2004
	Water - WDS	Greece	Kormas et al, 2010
	Water - WDS	USA	Marciano-Cabral et al, 2010
Shigella spp., *Salmonella* spp.	Biofilm - WDS	Germany	Schmeisser et al, 2003
	Water - WDS	Korea	Lee et al, 2006
Campylobacter spp.	Water - WDS	Finland	Hänninen et al, 2002
	Raw water	France	Gallay et al, 2006
Helicobacter pylori	Biofilm - WDS	England	Watson et al, 2004
	Biofilm - WDS	Portugal	Bragança et al, 2005
Legionella pneumophila	Biofilm - WDS	Germany	Emtiazi et al, 2004
	Water - WDS	The Netherlands	Diederen et al, 2007
	Biofilm - WDS	Portugal	Menaia et al, 2008
	Water - WDS	USA	Marciano-Cabral et al, 2010

Table 1. Pathogenic and opportunistic pathogenic bacteria detected in association with drinking water; WDS – water distribution systems, DWTP - drinking water treatment plant.

Experimental studies emphasized on bacteria ability of colonization, survival and multiplication in water associated biofilms, followed by dispersion in water phase in a planktonic state. The findings of Banning et al. (2003) suggested that the ability of *P. aeruginosa* to survive longer than *E. coli* in water associated biofilm could not be attributed to the association with the biofilm, rather than to the ability to utilize a wider range of organic molecules as carbon and energy sources compared to other *Enterobacteriaceae*. An increment in available nutrients may reduce *E. coli* survival in enhanced competition for nutrients and increased antagonism by the indigenous microbial population.

Lehtola et al. (2007) investigated the survival of faecal indices versus pathogenic bacteria and viruses, in drinking water biofilms experimentally infested with *E. coli, L. pneumophila, Mycobacterium avium* and canine calcivirus (as a surrogate for human norovirus). The results proved that pathogenic bacteria and virus particles entering water distribution systems can survive in biofilms for weeks, even in conditions of high-shear turbulent flow and may pose a risk to the consumers. Meanwhile, *E. coli* registered a limited survival to a few days in water and in biofilms, being a poor indicator of certain pathogens in biofilms. The study also showed that standard culture methods may seriously underestimate the real numbers of bacteria in water and biofilms.

Comparative evaluation of classical techniques involving bacterial growth on specific selective media and molecular methods based on 16s rDNA sequence identity reveals a high discrepancy between what was expected to grow and the species isolated from specific selective growth media. Bacterial analyses of water based on selective isolation and culturing approach is recommended to be interpreted with caution (September et al, 2007).

Experimental studies revealed also low detectable numbers by culture-based technique in case of potable water biofilms infected with *Campylobacter jejuni* (Lehtola et al, 2006). *C. jejuni* and *C. coli* waterborne epidemics registered in Finland (Hänninen et al, 2000) and France (Gallay et al, 2000) were associated to consumption of contaminated tap water with origin in polluted sources.

Severe outbreaks such as cholera caused by the ingestion of water contaminated with *Vibrio cholerae*, typhoid and paratyphoid enteric fevers caused by *Salmonella enterica* subsp. *enterica* serovar *Typhi*, respective serovar *Paratyphi*, shigellosis due to infections with *Shigella* species still occur in countries with insufficient access to safe water. But even in developed countries, outbreak events involving emerging pathogenic bacteria like *Legionella pneumophila*, waterborne E.coli O157:H7 and foodborne *E.coli* O104:H4 demonstrate the microbes' versatility and the fragility of humanity's victory over the nature.

2.1.2 Occurrence of protozoa in drinking water and associated biofilms

The food web in drinking water microbial consortia is based on heterotrophic bacteria, the next trophic level being represented by protozoa. Species of parasitic protozoa, including free living amoebae associated with infections in humans have been isolated from source waters and drinking water (table 2). Their presence represents a double threat to human health, being also related to amoeba-resisting bacteria, such as *Legionella* spp. and *Mycobacterium* spp., which proliferate in protozoa thus increasing the probability of causing diseases in humans (Marciano-Cabral et al, 2010).

Some protozoa, for example *Giardia* spp. and *Cryptosporidium* spp. may persist to hostile environment in drinking water, resist to different disinfection procedures and accumulate in biofilms under the form of cysts, respective oocysts. Experimental introducing *Cryptosporidium* oocysts for the prediction of behaviour in drinking water distribution system showed surface attachment and subsequent intermittent detachment, with exposure to high doses of chlorine (20mg/L) needed for the removal of substantial numbers of oocysts attached to pipe walls (Warneke et al, 2006). The study of Helmi et al. (2008) investigating the interaction of *Giardia lamblia* and *Cryptosporidium parvum* (oo)cysts in drinking water biofilms, revealed that protozoa are able to attach in biofilm matrix from the first day and survive extended periods of time, longer for *Cryptosporidium*. Viable (oo)cysts were recovered from biofilm and water phase for the whole period of investigation, of 34 days, turbulent shear stress influencing the detachment.

Protozoa	Samples type/Origin	Country	References
Flagellates: *Giardia lamblia*	Filtered water - DWTP Water - WDS Water - WDS Water - WDS	USA Canada Australia Spain	LeChevallier et al, 1991 Chung et al, 1998 Hellard et al, 2001 Carmena et al, 2007
Apicomplexa (Sporozoans): *Cryptosporidium parvum*	Filtered water - DWTP Water - WDS Water - WDS	USA Canada Spain	LeChevallier et al, 1991 Chung et al, 1998 Carmena et al, 2007
Amoebae: *Naegleria fowleri* *Acanthamoeba* spp. *Hartmannella* spp. *Vahlkampfia* spp.	Well water Biofilm, Water - WDS Biofilm, Water - WDS Water reservoir	USA USA USA Sudan	Blair et al, 2008 Marciano-Cabral et al, 2010 Shoff et al, 2010 Shanan et al, 2011

Table 2. Protozoa detected in raw water sources, water treatment plants (DWTP) and drinking water networks (WDS) and associated biofilms.

2.1.3 Occurrence of viruses in drinking water and associated biofilms

Sources of drinking water were investigated for the presence of enteric viruses, especially when gastrointestinal outbreaks occurred, and the results revealed episodes of faecal contamination in raw water. Epidemiological studies conducted supported the association between drinking water consumption and illness (table 3).

Viruses	Samples type/Origin	Country	References
Hepatitis A virus Hepatitis E virus	Well water Water - WDS	USA India	Bloch et al, 1990 Hazam et al, 2010
Noroviruses	Well water Spring water Groundwater Groundwater	USA Finland New Zeeland Korea	Parshionikar et al, 2003 Maunula et al, 2005 Hewitt et all, 2007 Koh et al, 2011
Coxsakie A viruses	Raw water	Taiwan	Hsu et al, 2009
Adenoviruses Rotaviruses	Drinking water sources	West Africa	Verheyen et al, 2009

Table 3. Viruses identified in raw water sources and water distribution networks (WDS).

The presence of enteric viruses associated with inadequate water supplies, poor sanitation and hygiene is mostly affecting developing countries (Ashbolt et al, 2004). Episodes of gastroenteritis caused especially by noroviruses attributed to contaminated drinking water have been reported also in developed countries. Inefficient raw water treatment and secondary contamination of distribution systems with sewage are of high concern, enteric viruses being generally more resistant than enteric bacteria to widely used free chlorine, chlorine dioxide and monochloramine disinfectants (LeChevallier & Au, 2004). Although no complete investigations regarding faecal indicators presence were performed in the considered studies (especially for intestinal enterococci as index of viruses), coliforms and *E. coli* have been detected in water samples in many cases of gastroenteritis outbreaks investigated (Parshionikar et al, 2003; Hewitt et al, 2007; Koh et al, 2011).

There is no evidence of viruses ability of multiplication in environmental biofilms, but they may survive for extended periods of time trapped in the matrix, similarly to protozoan (oo)cysts, and be detached in water column, where remain inactive until they find a host. Experimental studies using pilot scale systems demonstrated the ability of viruses to attach and accumulate into drinking water biofilms within one hour after inoculation, while their detachment in water phase is influenced by flow velocity (Lehtola et al, 2007; Helmi et al, 2008). The viral genomes were detected in biofilms over the whole period of both experiments (for 21, respectively 34 days). Helmi and co-workers, investigating the poliovirus infectivity, recovered the infectious viruses only for 6 days, when flow velocity increment from laminar to turbulent regimen was applied, concluding that detection of viral genome in biofilms is not sufficient to assess a risk associated with the presence of infectious particles.

2.1.4 Occurrence of fungi in drinking water and associated biofilms

Initially considered to be airborne, fungal infections in immunocompromised patients hospitalized in controlled atmospheric conditions raised the hypothesis of waterborne origin of aspergillosis (Anaissie & Costa, 2001). Opportunistic pathogens, potentially causing superficial or systemic infections, allergenic or toxigenic species of fungi (yeasts and moulds) have been isolated from drinking water worldwide, their presence being primary attributed to the ability of surfaces colonization as biofilms (table 4).

Fungi	Samples type/Origin	Country	References
Paenicillium spp.	Biofilm - WDS	USA	Doggett, 2000
Aspergillus spp.	Water - WDS	Germany	Göttlich et al, 2002
Cladosporium spp.	Biofilm, water - WDS	USA	Kelley et al, 2003
Epicoccum spp.	Water - WDS	Norway	Hageskal et al, 2006
Alternaria spp.	Water - WDS	Portugal	Gonçalves et al, 2006
Trichoderma spp.	Water WDS	Brazil	Pires-Gonçalves et al, 2008
Acremonium spp.	Water - WDS	Australia	Sammon et al, 2010
Exophiala spp.			
Phialophora spp.			
Fusarium spp.			
Mucor spp.			
Candida spp.			

Table 4. Fungi identified in drinking water distribution systems (WDS) and associated biofilms.

In some studies, the correlation with standard hygiene indicators was not found (Göttlich, 2002), other authors described negative correlations between bacteria and filamentous fungi, which may be explained either by competition for nutrients either by inhibiting toxins produced (Gonçalves et al, 2006) while in other investigations positive significant correlations were found between the presence of filamentous fungi, yeasts and bacteria in drinking water (Sammon et al, 2010). Regarding filamentous fungi behaviour in water distribution systems, deposition is attributed to highly resistant spores, while mycotoxins, taste and odour changing compounds producing implies germination and hyphal growth in biofilms. The occurrence of fungi in drinking water systems may have significant impact due to health effects of mycotoxins (such as aflatoxins): mutagenic, teratogenic, oestrogenic, carcinogenic and allergenic, although no reports of disease attributed to mycotoxins produced in the water distribution systems have been reported (Sonigo et al, 2011).

2.1.5 Occurrence of algae in drinking water and associated biofilms

Algae are assumed not to be characteristic to water distribution system biofilms due to the absence of light (Wingender and Flemming, 2011), but algal biomass is a major component of biofilms in surface source waters, water treatment and storage, in areas exposed to air and light. Experimental research designed by Chrisostomou et al (2009) emphasized on air-dispersed phytoplankton diversity and colonization potential of algal taxa in drinking water reservoir systems. Algal communities are associated to biofilms and may support bacterial growth, for example *Legionella* species (Declerck, 2010). Few recent studies investigating the presence of algae in drinking water are available (table 5).

Algae	Samples type/Origin	Country	References
Oocystis spp. *Xenococcus* spp.	Water - WDS	Spain	Codony et al, 2003
Anabaena spp. *Microcystis* spp. *Oscillatoria* spp. and many more	Water - WDS	Argentina	Ricardo et al, 2006
Microcystis aeruginosa *Chroococcus dispersus*	Water reservoirs	Greece	Lymperopoulou et al, 2011

Table 5. Algae identified in drinking water distribution systems (WDS).

Algal toxins, of which the most dangerous for humans is cyanobacterial microcystin, are considered chemical hazards in drinking water, especially when open-air reservoirs are used in water storage (Lymperopoulou et al, 2011). Algal growth and eutrophication in surface waters are widely investigated, with respect to ecological effects. In drinking water sources and throughout the water treatment process, distribution and storage, algal blooms raise issues about toxins releasing and aesthetic problems inducing, such as colour and smell. Algal removal in drinking water treatment is recommended to be carefully performed, in order not to disrupt the cells and release toxins in drinking water (LeChevallier and Au, 2004). Epidemiological studies are conducted worldwide in order to demonstrate the evidence of algal toxins in the environment and to evaluate their relatedness to illness in humans. Possible linkages between algae toxins in drinking water and health effects, including liver problems and diarrhoea in children were indicated by a survey in Namibia, although microcystin never exceeded the tolerable daily intake (Gunnarsson and Sanseovic, 2001).

3. Drinking water and associated biofilms – Chemical aspects

Detrimental effects of biofouling in drinking water distribution systems include chemical aspects, involving organic and inorganic compounds produced by the microorganisms inhabiting water phase, biofilms and sediments. Different volatile compounds, organic and inorganic acids, metal oxides and enzymes resulted in microbial metabolism or decay may cause aesthetic problems in water: colour, taste and odours and may also have an impact on the substratum, leading to microbially influenced corrosion.

3.1 Drinking water aesthetic problems

Aesthetic and organoleptic characteristics of water may be affected by a series of chemical substances, resulting in colour, odour and taste degradation. Such substances originate in microbial activity and decomposition in source waters and in distribution systems, disinfectants used in water treatment, materials used in pipes and joints in water networks. A list of these substances, related to microbial activity and decay that may be produced in the journey of drinking water from drinking water sources to the tap, that may influence consumers perception, is presented in table 6 (after the UK Environment Agency, 2004). These chemical compounds are usually attributed to microbial biofilms associated to drinking water processing and distribution.

Investigating the sources of taste and odour in drinking water in order to find their sources and mitigation strategies, Peter (2008) concluded that low concentrations in chlorine residuals, stagnant water, plastic pipes and particles accumulation in distribution systems may increase the generation of taste and odour compounds by favouring biofilm formation and microbial activity. Other sources of aesthetic problems in water may reside in the activity of bacteria involved in sulphur cycle, producing sulphur odours and yellow discoloration (US EPA, 2002). Oxidation and reduction of soluble metals may produce metal oxides, leading to consumer complaints about the metallic taste and yellow, black or brown staining water (Cerrato et al, 2006).

3.2 Microorganisms – Surface Interactions and microbially influenced corrosion

Biofouling proved to be interdependent on surface characteristics. Investigations of microbial reversible and irreversible attachment in primary or secondary colonization and in drinking water biofilms composition concluded as following:

- The hydrophobic/hydrophilic properties of the substrate are influencing biofilm formation. Exopolysacharides produced by some bacteria facilitate cell adhesion to hydrophilic surfaces, while exopolymers of other bacteria may show a preference for hydrophobic substrata (Beech et al, 2005).

Regarding the influence of the substratum on biofilm composition, copper materials appear to be colonized just by *L. pneumophila* in low numbers, inhibiting *P. aeruginosa* integration, while drinking water biofilms on elastomeric and polyethylene materials proved to be a better support for pseudomonads (Moritz et al, 2010).

- Pipe materials may be corroded, influencing disinfection effectiveness: corrosion products in iron pipes react with free chlorine and lead to residual disinfectants depletion. LeChevallier et al. (1987) detected high concentrations of coliforms only in tubercles formed on iron pipes and suggested few possible explanations: coliform growth stimulated by iron oxides; nutrient syntropy; sourface roughness; protection from disinfection. Iron pipes may be a better support for fungi also, when compared to PVC pipes (Doggett, 2000).

Drinking water flowing through PVC pipes contains three times the aqueous concentration of soluble manganese and 35 times the concentration of total manganese than present in the drinking water transported by iron pipes (Cerrato et al, 2006).

Microorganisms	Chemical substances produced	Aesthetic effects
Microbial decomposition	Indole, skatole, putrescine, cadaverine, β-phenylethylamine, butyric, propionic and stearic acids	Fishy, grassy, woody tastes Faecal, rotten, cheese, pungent odours
Algae decomposition	Mercaptan, dimethyl sulphide, polysulphides	Fishy, swampy, septic odours
Algae decomposition/activity	n-hexanal, n-heptanal, isomers of decadienal sulphur compounds, terpenes, aromatic compounds, esters	Fishy odours Rotten eggs odours Aromatic odours
Pseudomonas spp. Flavobacterium spp. Aeromonas spp. Paenicillium caeseicolum	Dimethyl polysulphides	Swampy odours
Fungi Chaetomium globosum Basidiobolus ranarumi Actinomycetes	Geosmin Cadin-4-ene-1-ol 2-isopropyl-3-methoxypyrazine	Earthy, musty taste and odour Woody, earthy odour Musty, mouldy potato odour
Actinomycetes: Streptomyces spp. Nocardia spp. Microbiospora spp Cyanobacteria: Anabaena spp. Microcystis spp. Oscillatoria spp. Aphanizomenon spp. Algae: Chlorophyceae Bacillariophyceae	Geosmin 2-methylsorboneol	Earthy, musty taste and odours
Sulphur oxidizing/reducing bacteria Sulphate reducing bacteria	Sulphuric acid, sulphates, sulphur, methyl mercaptan, hydrogen sulphide Metal sulphides (ferrous sulphide)	Rotten eggs, rotten cabbage odours Yellow, brown, black staining
Metals oxidizing/ reducing bacteria	Metal oxides	Rusty or metallic taste Brown, black staining

Table 6. Chemical compounds produced by microbial decomposition and metabolism, affecting taste and odour of drinking water.

Microbially influenced corrosion represents another undesirable impact of biofilms associated to drinking water treatment and distribution, involving metallic or non-metallic materials deterioration as a result of pipes inner surface biofouling.

Physiological groups of bacteria classified on account of the ability to use different substrates in their nutrition or in respiration are summarized in table 7 (Drăgan-Bularda & Kiss, 1986; Drăgan-Bularda & Samuel, 2006; Muntean, 2009). Their representative species may belong to microbial communities of source waters and sediments, enter drinking water treatment plants and distribution systems in a planktonic state and adhere to surfaces or become members of established biofilms (Costerton, 1994). Their metabolites have significant impacts on drinking water quality, either being released in bulk water where may react with other compounds, for example with disinfectants, leading to toxic disinfection-by products (as trihalomethanes), or by remaining in biofilm matrix where acting upon pipes surfaces and inducing corrosion.

Physiologycal groups of bacteria	Representatives	Metabolites produced
Ammonifying bacteria	*Bacillus* spp. *Clostridium* spp. *Pseudomonas* spp. *Burkholderia* spp.	Ammonium Ammonia
Nitrosifiers (Ammonia oxidizing bacteria)	*Nitrosomonas* spp. *Nitrocystis* spp. *Nitrospira* spp. *Nitrosolobus* spp. *Nitrosovibrio* spp.	Nitrite ions
Nitrifying bacteria (Nitrite oxidizing bacteria)	*Nitrobacter* spp *Nitrococcus* spp. *Nitrospira* spp. *Nitrospina* spp.	Nitrate ions
Denitrifying bacteria	*Paracoccus denitrificans* *Pseudomonas stutzeri* *Thiobacillus denitrificans* *Alcaligenes* spp. *Bacillus* spp.	Nitrous oxide Nitrogen
Sulphur reducing bacteria	*Desulfuromonas* spp. *Proteus* spp.	Hydrogen sulphide
Sulphate reducing bacteria	*Desulfovibrio desulfuricans* *Desulfovibrio sulfodismutans* *Desulfotomaculum* spp. *Desulfonema* spp. *Desulfosarcina* spp. *Desulfobacter* spp. *Desulfococcus* spp. *Desulfomicrobium* spp.	Hydrogen sulphide
Sulphur oxidizing bacteria:	*Thiobacillus* spp. *Sulfolobus* spp. *Beggiatoa* spp. *Thiothrix* spp.	Sulphuric acid Sulphates Sulphur
Iron reducing bacteria	*Sphaerotilus natans* *Leptothrix ochracea* *Crenothrix polyspora*	Iron (Fe^{2+}) oxides
Iron oxidizing bacteria	*Galionella feruginea* *Ferrobacillus ferooxidans* *Thiobacillus ferooxidans*	Iron (Fe^{3+}) oxides
Manganese oxidizing/reducing bacteria	*Sphaerotilus discophorus* *Pseudomonas* spp. *Metallogenium* spp. *Pedomicrobium* spp. *Bacillus* spp. *Micrococcus* spp. *Vibrio* spp.	Manganese oxides

Table 7. Physiological groups of bacteria, their representatives and metabolites.

Chemical and enzymatic microbial products resulted in biofilms activity may induce corrosion and related effects by different mechanisms:

- oxygen concentration cells and anaerobic sites generation (promoting growth of anaerobic bacteria);
- formation of iron concentration cells by the activity of iron and manganese oxidizing bacteria;
- metabolites such as acids produced by bacteria have corrosive action upon the surface;
- production of depolarizing enzymes within the biofilm matrix, which may persist longer than viable cells;
- exopolymers produced by slime forming bacteria stimulate biofilm formation and biomass accumulation;
- the binding capacity of biofilm matrix which may lead to deposits accumulation with clogging effects (Beech et al, 2005; Coetser and Cloete, 2005).

Some of the recommended strategies in drinking water associated biofilm control are: source waters protection, appropriate treatment, infrastructure contamination prevention, pipes and reservoirs maintenance, corrosion control, appropriate disinfection practices, nutrient levels reducing, water quality monitoring, personnel training, water safety plans implementation.

We are still living in an age of surfaces, even the remark was first said by Oscar Wild's character in 1895. Having in mind the virtual idea of self-cleaning surfaces, researchers in nanotechnology field are targeting innovative repellent materials with a wide range of applications, for the biofouling control. The superhydrophobicity models such as "the lotus effect" characterizing the lotus (*Nelumbo nucifera*) leaf, offered by natural patterns are investigated at nanoscale. The interdependence between surface roughness, reduced particle adhesion and water repellence proved to be the keystone in the self-cleaning mechanism of many biological surfaces (Barthlott & Neinhuis, 1997).

4. Conclusions

The present review emphasize on the following recent and relevant findings:

- Biofilms associated with drinking water are ubiquitous, harbouring bacterial pathogens, opportunistic pathogens, parasitic protozoa, viruses, toxins releasing fungi and algae;
- Microbial consortia in contact with drinking water have significant impacts upon water quality and may threat human health when contamination events occur;
- Access to safe water continues to be a target for developing countries, unfulfilled at the moment;
- Even in developed countries, where substantial efforts are submitted in order to ensure population's access to a high quality drinking water, microbial versatility represents an endless source of problems, with respect to opportunistic pathogens emergence;
- Microbial communities in water networks and biofilms represent complex ecosystems; their ecology is influenced by a series of abiotic and biotic factors: raw water sources quality, temperature, flow rate and system hydraulics, nutrient concentration, pipe material, particles accumulation, ingress and intrusion, water treatment, water disinfection and microbial interactions;
- Further research is needed in order to understand attached microbial consortia for biofouling prevention and control in drinking water industry, as a matter of public security.

5. References

Alam, M.; Sultana, M.; Nair, G.B.; Siddique, A.K.; Hasan, N.A.; Sack, R.B.; Sack, D.A.; Ahmed, K.U.; Sadique, A.; Watanabe, H.; Grim, C.J.; Huq, A. & Colwell, R.R. (2007). Viable but nonculturable *Vibrio cholerae* O1 in biofilms in the aquatic environment and their role in cholera transmission. *Proceeding of the National Academy of Sciences of the USA*, Vol.4, No.5, pp. 17801-17806, www.pnas.org/cgi/content/full/0705599104/DC1

Allen, M.J. (2011). *Escherichia coli* – the most relevant microbial public health indicator for drinking water. *Faecal indicators: problem or solution?* Conference presentation, Edinburgh, UK

Anaissie, E.J. & Costa, S.F. (2001). Nosocomial aspergillosis is waterborne. *Clin Infect Dis*, Vol.33, pp. 1546-1548

Ashbolt, N.J. (2004). Microbial contamination of drinking water and disease outcomes in developing countries. *Toxicology*, Vol.198, pp. 229-238

Barthlott, W. & Neinhius, C. (1997). Purity of the sacred lotus, or escape from contamination in biological surfaces. *Planta*, Vol.202, pp. 1-8

Banning, N.; Toze, S. & Mee, B.J. Persistence of biofilm-associated *Escherichia coli* and *Pseudomonas aeruginosa* in groundwater and treated effluent in a laboratory model system. *Microbiology*, Vol.149, pp. 47-55

Beech, I.B.; Sunner, J.A. & Hiraoka, K. (2005). Microbe-surface interactions in biofouling and biocorrosion processes. *Int Microbiol*, Vol.8, pp. 157-168

Blair, B; Sarkar, P.; Bright, K.R.; Marciano-Cabral, F. & Gerba,C.P. (2008). *Naegleria fowleri* in well water. *Emerg Infect Dis*, Vol.14, No.9, pp. 1499-1500

Bloch, A.B.; Stramer, S.L.; Smith, J.D.; Margolis, H.S.; Fields, H.A.; McKinley, T.W.; Gerba, C.P.; Maynard, J.E. & Sikes, R.K. (1990). Recovery of hepatitis A virus from a water supply responsible for a common source outbreak of hepatitis A. *Am J Public Health*, Vol.80, No.4, pp. 428-430

Bomo, A.M.; Storey, M.V. & Ashbolt, N.J. (2004). Detection, integration and persistence of aeromonads in water distribution pipe biofilm. *J Water Health*, Vol.2, No.2, pp. 83-96

Bragança, S.M.; Azevedo, N.F.; Simões, L.C.; Vieira, M.J. & Keevil, C.W. (2005). Detection of *H. pylori* in biofilms formed in a real drinking water distribution system using peptide nucleic acid fluorescence in situ hybridization. *Biofilm Club*, pp. 231-239

Carmena, D.; Aguinagalde, X.; Ziggoraga, C.; Fernández-Crespo, J.C. & Ocio, J.A. (2007). Presence of *Giardia* cysts and *Cryptosporidium* oocysts in drinking water supplies in northern Spain. *J Appl Microbiol*, Vol.102, pp. 619-629

Cerrato, J.M.; Reyes, L.P.; Alvarado, C.N. & Dietrich, A.M. (2006). Effect of PVC and iron materials on Mn(II) deposition in drinking water distribution systems. *Water Res*, Vol.40, pp. 2720-2726

Chambless, J.D.; Hunt, S.M. & Stewart, P.S. (2005). A three-dimensional computer model of four hypothetical mechanisms protecting biofilms from antimicrobials. *Appl Environ Microbiol*, Vol.72, No.3, pp. 2005-2013

Characklis, W.G. (1988). *Bacterial regrowth in distribution systems*, Project Report, American Water Works Association, USA, http://waterrf.org/ProjectsReports/Executive SummaryLibrary/90532_102_profiles.pdf

Chauret, C.; Volk, C.; Creason, R.; Jarosh, J.; Robinson, J. & Wrnes, C. (2001). Detection of *Aeromonas hydrophila* in a drinking-water distribution system: a field and pilot study. *Can J Microbiol*, Vol.47, No.8, pp. 782-786

Chrisostomou, A.; Moustaka-Gouni, M.; Sgardelis, S. & Lanaras, T. (2009). Air-dispersed phytoplankton in a Mediterranean river-reservoir system (Aliakmon-Polyphytos, Greece). *J Plankton Res*, Vol.31, pp. 877-884

Chung, E.; Aldom, J.E.; Chagla, A.H.; Kostrzynska, M.; Palmateer, G.; Trevors, J.T.; Unger, S. & De Grandis, S. (1998). Detection of *Cryptosporidium parvum* oocysts in municipal water samples by the polymerase chain reaction. *J Microbiol Meth*, Vol.33, No.2, pp. 171-180

Codony, F.; Miranda, A.M. & Mas, J. (2003). Persistence and proliferation of some unicellular algae in drinking water systems as a result of their heterotrophic metabolism. *Water SA*, Vol.29, No.1, pp. 113-116, http://www.wrc.org.za

Coetser, S.E. & Cloete, T.E. (2005). Biofouling and biocorrosion in industrial water systems. *Crit Rev Microbiol*, Vol.31, No.4, pp. 213-232

Costerton, J.W.; Cheng, K.J.; Gessey, G.G.; Ladd, T.I.; Nickel, J.C.; Dasgupta, M. & Marrie, T.J. (1987). Bacterial biofilms in nature and disease. *Annu Rev Microbiol*, Vol.41, pp. 435-464

Costerton, J.W. (1994). Structure of biofilms, In: *Biofouling and biocorrosion in industrial water systems*, Geesey, G.G.; Lewandowski, Z. & Flemming, H.C. (Ed.) CRC Press, ISBN 0 87371 928 X, USA

Declerck, P. (2010). Biofilms: the environmental playground of *Legionella pneumophila*. *Environ Microbiol*, Vol.12, No.3, pp. 557-566

Diederen, B.M.W.; de Jong, C.M.A. & Aarts, I. (2007). Molecular evidence for the ubiquitous presence of *Legionella* species in Dutch tap water installations. *J Water Health*, Vol.5, No.3, pp. 375-383

Doggett, M.S. (2000). Characterization of fungal biofilms within a municipal water distribution system. *Appl Environ Microb*, Vol. 66, No. 3, pp. 1249-1251

Drăgan-Bularda, M. & Kiss, S. (1986). *Soil microbiology* (in Romanian), Babeş-Bolyai University of Cluj-Napoca Press, Romania

Drăgan-Bularda, M. & Samuel, A.D. (2006). *General microbiology* (in Romanian), University of Oradea Press, ISBN 10 973 759 056 2

Emtiazi, F.; Schwartz, T.; Marten, S.M.; Krolla-Sidenstein, P. & Obst, U. (2004). Investigation of natural biofilms formed during the production of drinking water from surface water embankment filtration. *Water Res*, Vol.38, pp. 1197-1206

Exner, M.; Vacata, V. & Gebel, J. (2003). Public health aspects of the role of HPC – an introduction, In: *Heterotrophic plate counts and drinking water safety. The significance of HPCs for water quality and human health,* Bartram, J.; Cotruvo, J.; Exner, M.; Fricker, C. & Glasmacher, A., (Ed.), IWA Publishing, ISBN 1 84339 025 6, London, UK, pp. 12-19

Farkas, A.; Drăgan-Bularda, M.; Ciataràş, D.; Bocoş, B. & Ţigan, Ş. (2011): Opportunistic pathogens and faecal indicators assessment in drinking water associated biofilms in Cluj, Romania (in preparation)

Fields, B.S.; Benson, R.F. & Besser, R.E. (2002). *Legionella* and Legionnaire's disease: 25 years of investigation. *Clin Microbiol Rev*, Vol.15, No.3, pp. 506-526

Flemming, H.C.; Percival, S.L. & Walker, J.T. (2002). Contamination potential of biofilms in water distribution systems. *Water Supp*, Vol.2, No.1, pp. 271-280

Flemming, H.C. (2008). Why microorganisms live in biofilms and the problem of biofouling. *Biofouling*, pp. 3-12, doi: 10.1007/7142

Gallay, A.; De Valk, H.; Cournot, M.; Ladeuil, B.; Hemery, C.; Castor, C.; Bon, F.; Mégraud, F.; Le Cann, P. & Desenclos, J.C. (2006). A large multi-pathogen waterborne community outbreak linked to faecal contamination of a groundwater system, France, 2000. *Clin Microbiol Infec*, Vol.12, No.6, pp 561-570

Gavriel, A.A.; Landre, J.P.B. & Lamb, A.J. (1998). Incidence of mesophilic *Aeromonas* within a public drinking water supply in north-east Scotland. *J Appl Microbiol*, Vol.84, pp. 383-392

Gillings, M.R.; Xuejun, D.; Hardwick, S.A. & Holley, M.P. (2009). Gene cassettes encoding resistance to quaternary ammonium compounds: a role in the origin of clinical class 1 integrons? *The ISME Journal*, Vol. 3, pp. 209-215

Glasmacher, A.; Engelhart, S. & Exner, M. (2003). Infections from HPC organisms in drinking-water amongst the immunocompromised, In: *Heterotrophic Plate Counts and Drinking-water Safety*, Bartram, J.; Cotruvo, J.; Exner, M.; Fricker, C. & Glasmacher, A., (Ed.), IWA Publishing, ISBN 1 84339 025 6, London, UK, pp. 137-145

Gonçalves, A.B.; Paterson, R.M. & Lima, N. (2006). Survey and significance of filamentous fungi from tap water. *Int J Hyg Environ Health*, Vol.209, pp. 257-264

Göttlich, E.; van der Lubbe, W.; Lange, B.; Fiedler, S.; Melchert, I.; Reifenrath, M.; Flemming, H.C. & Hoog, S. (2002). Fungal flora in groundwater-derived public drinking water. *Int J Hyg Environ Health*, Vol.205, No.4, pp. 269-279

Gunnarsson, H. & Sanseovic, A.M. (2001). *Possible linkages between algae toxins in drinking water and related illnesses in Windhoek, Namibia*, Thesis, Kristianstad University, Sweden, hkr.divaportal.org/smash/get/diva2:231226/FULLTEXT01

Hageskal, G.; Knutsen, A.K.; Gaustad, P.; de Hoog, G.S. & Skaar, I. (2006). Diversity and significance of mold species in Norwegian drinking water. *Appl Environ Microbiol*, Vol.72, No.12, pp. 7586-7593

Hazam, R.K.; Singla, R.; Kishore, J.; Singh, S.; Gupta, R.K. & Kar, P. (2010). Surveillance of hepatitis E virus in sewage and drinking water in a resettlement colony of Delhi: what has been the experience? *Arch Virol*, Vol.155, No.8, pp. 1227-1233

Hänninen, M.L.; Haajanen, H.; Pummi, T.; Wermundsen, K.; Katila, M.L.; Sarkkinen, H.; Miettinen, I. & Rautelin, H (2003). Detection and typing of *Campylobacter jejuni* and *Campylobacter coli* and analysis of indicator organisms in three waterborne outbreaks in Finland. *Appl Environ Microb*, Vol.69, No.3, pp. 1391-1396

Hellard, M.E.; Sinclair, M.I.; Forbes, A.B. & Fairley, C.K. (2001). A randomized, blinded, controlled trial investigating the gastrointestinal health effects of drinking water quality. *Environ Health Persp*, Vol.109, No.8, pp. 773-778

Helmi, K.; Skraber, S.; Gantzer, C.; Williame, R.; Hoffmnn, L. & Cauchie, H.M. (2008). Interactions of *Cryptosporidium parvum*, *Giardia lamblia*, vaccinal poliovirus type 1 and bacteriophages φX174 and MS2 with a drinking water biofilm and a wastewater biofilm. *Appl Environ Microb*, Vol.74, No.7, pp. 2079-2088

Hewitt, J.; Bell, D.; Simmons, G.C.; Rivera-Aban, M.; Wolf, S. & Greening, G.E. (2007). Gastroenteritis outbreak caused by waterborne Norovirus at a New Zeeland ski resort. *Appl Environ Microb*, Vol.73, No.24, pp. 7853-7857

Hsu, B.M.; Chen, C.H.; Wan, M.T.; Chang, P.J & Fan, C.W. (2009). Detection and identification of enteroviruses from various drinking water sources in Taiwan. *J Hydrol*, Vol.365, No.1-2, pp. 134-139

Hunter, P.R.; Payment, P.; Ashbolt, N & Bartram, J. (2003). Assessment of risk, In: *Assessing microbial safety of drinking water*, IWA Publishing, ISBN 1 84339 036 1, London, UK

Juhna, T.; Birzniece, D.; Larsson, S.; Zulenkovs, D.; Sharipo, A.; Azevedo, N.F.; Menard-Szczebara, F.; Castagnet, S.; Feliers, C. & Keevil, C.W. (2007). Detection of *Escherichia coli* in biofilms from pipe samples and coupons in drinking water distribution networks. *Appl Environ Microb* Vol.73, No.22, pp. 7456–7464

Kelley, J.; Kinsey, G.; Peterson, R. & Brayford, D. (2003). Identification and control of fungi in distribution systems. *Water Quality Technology Conference Proceedings*, Awwa Research Foundation Denver, USA, www.biosan.com/pubs/Identification Significance.pdf

Koh, S.J.; Cho, H.G.; Kim, B.H. & Choi, B.Y. (2011). An outbreak of gastroenteritis caused by Norovirus-contaminated groundwater at a waterpark in Korea. *Infectious Diseases, Microbiology & Parasitology*, Nr.26, pp. 28-32

Kormas, K.A.; Neofitou, C.; Pachiadaki, M. & Koufostathi, E. (2010). Changes of the bacterial assemblages throughout an urban drinking water distribution system. *Environ Monit Assess*, No.165, pp. 27-38

Latimer, J.; McLeod, C.; Jackson, R.; Bunch, J.; Graham, A.; Stokes S. & Poole, R. (2010). *Escherichia coli* biofilms: gene expression and elemental heterogeneity. *Biofilms 4 Conference Handbook*, Winchester, UK, pp. 86

LeChevallier, M.W.; Babcock, T.M. & Lee, R.G. (1987). Examination and characterization of distribution system biofilms. *Appl Environ Microb*, Vol.53, No.12, pp. 2714-2724

LeChevallier, M.W.; Norton, W.D. & Lee, R.G. (1991). *Giardia* and *Cryptosporidium* in filtered drinking water supplies. *Appl Environ Microb*, Vol.57, No.9, pp. 2617-2621

LeChevallier, M.W. & Au, K.K. (2004). *Water treatment and pathogen control*, IWA Publishing, ISBN 1 84339069 8, London, UK

Lee, D.G.; Kim, S.J. & Park, S.J. (2006). Effect of reservoirs on microbiological water qualities in a drinking water distribution system. *J Microbiol Biotechn*, Vol.16, pp. 1060-1067

Lehtola, M.J.; Pitkänen, T.; Miebach, L. & Miettinen, I.T. (2006). Survival of *Campylobacter jejuni* in potable water biofilms: a comparative study with different detection methods. *Water Sci Technol*, Vol.54, No.3, pp. 57-61

Lehtola, M.J.; Torvinen,E.; Kusnetsov, J.; Pitkänen, T.; Maunula,L.; von Bonsdorff, C.H.; Martikainen, P.J.; Wilks, S.A.; Keevil, C.W & Miettinen, I.T. (2007). Survival of *Mycobacterium avium*, *Legionella pneumophila*, *Escherichia coli*, and calciviruses in drinking water-associated biofilms grown under high-shear turbulent flow. *Appl Environ Microb*, Vol.73, No.9, pp. 2854-2859

Lymperopoulou, D.S.; Kormas, K.A.; Moustaka-Gouni, M. & Karagouni, A.D. (2011). Diversity of cyanobacterial phylotypes in a Mediterranean drinking water reservoir (Marathonas, Greece). *Environ Monit Assess*, Vol.173, pp.155-165

Marciano-Cabral, F.; Jamerson, M. & Kaneshiro, E.S. (2010). Free-living amoebae, *Legionella* and *Mycobacterium* in tap water supplied by a municipal drinking water utility in the USA. *J Water Health*, Vol.8, No.1, pp. 71-82

Martiny, A.C.; Jørgensen, T.M.; Albrechtsen, H.J.; Arvin, E. & Molin, S. (2003). Long term succession of structure and diversity of a biofilm formed in a model drinking water distribution system. *Appl Environ Microb*, Vol.69, No.11, pp. 6899-6907

Maunula, L.; Miettinen, I.T. & von Bonsdorff, C.H. (2005). Norovirus outbreaks from drinking water. *Emerg Infect Dis*, Vol.11, No.11, pp. 1716-1721

Menaia, J.; Benoliel, M.; Lopes, A.; Neto, C.; Ferreira, E.; Mesquita, E.& Paiva, J. (2008). Assessment of Lisbon drinking water distribution network biofilm colonization and associated hazards. *Wa Sci Technol*, Vol.8, No.4, pp. 421-426

Moritz, M.M.; Flemming, H.C. & Wingender, J. (2010). Integration of *Pseudomonas aeruginosa* and *Legionella pneumophila* in drinking water biofilms grown on domestic plumbing materials. *Int J Hyg Environ Health*, Vol.213, No.3, pp. 190-197

Muntean, V. (2009). *General microbiology* (in Romanian), Babeş-Bolyai University of Cluj-Napoca Press, Romania

Murga, R.; Forster, T.S.; Brown, E.; Pruckler, J.M.; Fields, B.S. & Donlan, R.M. (2001). Role of biofilms in the survival of *Legionella pneumophila* in a model potable-water system. *Microbiology*, Vol.147, pp. 3121-3126

Parshionikar, S.U.; William-True, S.; Fout, G.S.; Robbins, D.E.; Seys, S.A.; Cassady, J.D. & Harris, R. (2003). Waterborne outbreak of gastroenteritis associated with a Norovirus. *Appl Environ Microb*. Vol. 69, No.9, pp. 5263-5268

Payment, P. & Robertson, W. (2004). The microbiology of piped distribution systems and public health, In: *Safe piped water: Managing Microbial Water Quality in Piped Distribution Systems*, Ainsworth, R. (Ed.), IWA Publishing, ISBN 1 84339 039 6, London, UK

Payment, P. & Locas, A. (2011). Pathogens in water: value and limits of correlation with microbial indicators. *Ground Water*, Vol.49, No.1, pp 4-11

Peter, A. (2008). *Taste and odor in drinking water: Sources and mitigation strategies*, PhD Thesis, Swiss Federal Institute of Technology Zurich, Switzerland, http://e-collection. ethbib.ethz.ch/eserv/eth:30628/eth-30628-02.pdf

Pires-Gonçalves, R.H.; Sartori, F.G.; Montanari, L.B.; Zaia, J.E.; Melhem, M.S.C.; Mendes-Giannini, M.J.S. & Martins C.H.G. (2008). Occurrence of fungi in water used at a haemodialysis centre. *Lett Appl Microbiol*, Vol.46, pp. 542-547

Razzolini, M.T.P.; Bari, M.; Sanchez, P.S. & Sato, M.I.Z. (2008). *Aeromonas* detection and their toxins in drinking water from reservoirs and drinking fountains. *J Water Health*, Vol.6, No.1, pp. 117-123

Reynolds, K.A.; Mena, K.D. & Gerba, C.P. (2008). Risk of waterborne illness via drinking water in the United States. *Rev Environ Contamin T*, Vol.192, pp. 117-158

Ricardo, E.; Leda, G. & Luis, F. (2006). Drinking water: problems related to water supply in Bahia Blanca, Argentina. *Acta Toxicol Argent*, Vol.14, No.2, pp. 23-30.

Riley, M.R.; Gerba, C.P. & Elimelech, M. (2011). Biological approaches for addressing the grand challenge of providing access to clean drinking water. *J Biol Eng*, Vol.5, No.2

Rusin, P.A.; Rose, J.B.; Haas, C.N. & Gerba, C.P. (1997). Risk assessment of opportunistic bacterial pathogens in drinking water. *Rev Environ Contam T*, Vol.152, pp. 57-83

Sammon, N.B.; Harrower, K.M.; Fabbro, L.D. & Reed, R.H. (2010). Incidence and distribution of microfungi in a treated municipal water supply system in sub-tropical Australia. *Int J Environ Res Public Health*, Vol.7, pp. 1597-1611

Schmeisser, C.; Stockigt, C.; Raasch, C.; Wingender, J.; Timmis, K.N.; Wenderoth, D.F.; Flemming, H.C.; Liesegang, H.; Schmitz, A.; Jaeger, K.E. & Streit, W.R. (2003). Metagenome survey of biofilms in drinking water networks. *Appl Environ Microb*, Vol.69, No.12, pp. 7298-7309

September, S.M.; Els, F.A.; Venter, S.N. & Brozel, V.S. (2007). Prevalence of bacterial pathogens in biofilms of drinking water distribution systems. *J Water Health*, Vol.5, No.2, pp. 219-227

Servais, P.; Anzil, A.; Gatel, D. & Cavard, J. (2004). Biofilm in the Parisian suburbs drinking water distribution system. *J Water Supply Res T*, Vol.53, pp. 313-324

Shanan, S.; Abd, H.; Hedenström, I.; Saeed, A. & Sandström, G. (2011). Detection of *Vibrio cholerae* and *Acanthamoeba* species from same natural water samples collected from different cholera endemic areas in Sudan. *BioMed Central Research Notes*, Vol.4, No.109, http://www.biomedcentral.com/1756-0500/4/109

Shoff, M.E.; Rogerson, A.; Kessler, K.; Schatz, S. & Seal, D.V. (2008). Prevalence of *Acanthamoeba* and other naked amoebae in South Florida domestic water. *J Water Health*, Vol.6, No.1, pp. 99-104

Sonigo, P.; De Toni, A. & Reilly, K. (2011). *A review of fungi in drinking water and the implications for human health*, Report, Bio Intelligent Service, France

Srinivasan, S. (2008). *Managing bacterial regrowth and presence in drinking water distribution systems*, Ph.D. thesis, University of Wisconsin-Madison, USA

Szewzik, U.; Szewzyk, R.; Manz, W. & Schleifer, K.H. (2000). Microbiological safety of drinking water. *Annu Rev Microbiol*, Vol.54, pp. 81-127

UK Environment Agency (2004). *The microbiology of drinking water (2004) - Part 11 – Taste, odour and related aesthetic problems*, London, UK, http://www.environment-agency.gov.uk/static/documents/Research/mdwpart112004_859972.pdf

US Environmental Protection Agency (1992). *Control of biofilm growth in drinking water distribution systems*, Seminar publication, Washington, USA, www.epa.gov/nrmrl/pubs/625r92001/625r92001.html

van der Kooij, D. (2003). Managing regrowth in drinking water distribution systems, In: *Heterotrophic plate counts and drinking water safety. The significance of HPCs for water quality and human health*, Bartram, J.; Cotruvo, J.; Exner, M.; Fricker, C. & Glasmacher, A., (Ed.), IWA Publishing, ISBN 1 84339 025 6, London, UK, pp. 199-232

Verheyen, J.; Timmen-Wego, M.; Laudien, R.; Boussaad, I.; Sen, S.; Koc, A.; Uesbeck, A.; Mazou, F. & Pfister, H. (2009). Detection of adenoviruses and rotaviruses in drinking water sources used in rural areas of Benin, West Africa. *Appl Environ Microb*, Vol.75, No.9, pp. 2798-2810

Warneke, M. (2006). *Cryptosporidium oocyst interactions with drinking water pipe biofilm*, Research report, Cooperative Research Center for Water Quality and Treatment, Adelaide, Australia, www.wqra.com.au/publications/document-search/ ?download=30

Watkins, J. & Jian, X. (1997). Cultural methods of detection for microorganisms: recent advances and successes. In: *The microbiological quality of water*, Sutcliffe, D.W (Ed.), Freshwater Biological Association, ISBN 0 900386 57 6, Ambleside, UK, pp 19-27

Watson, C.L.; Owen, R.J.; Said, B.; Lai, S.; Lee, J.V.; Surman-Lee, S. & Nichols, G. (2004). Detection of *Helicobacter pylori* but not culture in water and biofilm samples from drinking water distribution systems in England. *J Appl Microb*, No.97, pp. 690-698

White, D.C.; Kirkegaard, R.D.; Palmer, R.J.; Flemming, C.A.; Chen, G.; Leung, K.T.; Phiefer, C.B. & Arrage, A.A., (1999). The biofilm ecology of microbial biofouling, biocide resistance and corrosion, In: *Biofilms in the aquatic environment*, Keevil, C.W.; Godfree, A.; Holt, D. & Dow, C., (Ed.), The Royal Society of Chemistry, ISBN 0 85404 758 1, Cambridge, UK, pp. 120-130

Wingender, J. & Flemming, H.C. (2011). Biofilms in drinking water and their role as reservoir for pathogens. *Int J Hyg Environ Heal*, Vol.213, pp. 190-197

5

Water Quality After Application of Pig Slurry

Radovan Kopp

Mendel University in Brno, Department of Fisheries and Hydrobiology
Czech Republic

1. Introduction

Pig slurry is a complex organic-mineral fertilizer with high fertilising efficiency and it is comparable with farmyard manure. The pig house with excremental ending and subsequent use of slurry is very popular in agriculturally advanced countries because of economic efficiency, work culture and hygiene. Pig house is operationally by 30 - 40 % cheaper than using bedding. Properly made and treated slurry means a significant source of organic matter, nutrients, bacteria and stimulators which if correctly applied, increase fertility and provide a significant cost saving. These slurry nutrients are easily usable for phytoplankton. Nitrogen is found mainly in inorganic form (50 – 60%), just 10% is nitrate and rest forms organic nitrogen. Phosphorus is bound to organic matter and potassium is included in the urine. Chemically the high fertilising efficiency of slurry depends on C:N ratio, which is usually 4-8:1. This ratio affects the mineralization of organic matter, release of N from organic fixation and energy utilization for multiplication of microorganisms. Composition of the slurry is in the Table No.1.

Dry matter	pH	Conductivity	ANC	COD_{Cr}	Ca^{2+}
%		$mS.m^{-1}$	$mmol.l^{-1}$	$g.l^{-1}$	$mg.l^{-1}$
1.4 - 7.5	6.21 – 7.90	1918 - 2715	179 - 227	16.0 - 18.5	300 - 1300
TP	TN	$N-NH_4$	$N-NO_3$	$N-NO_2$	K^+
$mg.l^{-1}$	$mg.l^{-1}$	$mg.l^{-1}$	$mg.l^{-1}$	$mg.l^{-1}$	$mg.l^{-1}$
33 - 2200	1440 - 6000	1000 - 4348	16.9 – 36.2	1.28 – 3.34	315 - 7686

Table 1. Values of physical and chemical parameters in pig slurry applied into ponds. (min-max interval, amended by more authors), ANC - acid neutralization capacity, COD - chemical oxygen demand, TP - total phosphorus, TN – total nitrogen

An additional fertilizer in intensively managed fish-ponds usually does not mean that significant increase of fish production as observed in nutrient-poor waters. Moreover, excessive application of fertilizer may lead to environmental pollution and sanitary problems. Fairly high fish production can be achieved by organic fertilizers. The application of pig slurry to ponds as a fertilizer is widely used in many countries in order to increase plankton production and fish growth. That's why manuring is considered to be a cheap way to increase carp production in the pond. Some fish species also feed directly upon these waste, which is enriched by microorganisms with high protein value. Further, different types of manure can be also added into the low-level protein feeds. The type of pig slurry, quality, quantity and the season affects water quality and production of the pond and finally on well-being and growth of fish. Additionally, recycling of pig slurry could be interesting for fish farming in order to reduce the impact of intensive pig farming on environment.

Pig slurry applied to ponds as organic fertilizer necessary needs a decomposition before their nutritional contents are released, assimilated and utilized by plankton. The nutrient content of pig slurry may vary with time, and nutrient availability to phytoplankton growth remains unclear. The rate of nutrients released from animal manure is an important factor to regulate the frequency and amount of manure required to fertilize fish ponds. Pig slurry is destined mainly for refilling of carbon into water and modification of proportion basic biogenic elements (C, N, P). During the water treatment by slurry the amount of carbon dioxide in pond was relatively higher (Fig. 1). After the last application of pig slurry in early may in 2001 alkalinity increased and value of pH decreased which caused a significant increased in values of TIC due to the weather conditions and lower intensity of photosynthesis. On the other hand another TIC value decrease occurred in 2002 (at the end of March) due to higher phytoplankton development. This expansion decreased carbonate content, alkalinity and increased pH.

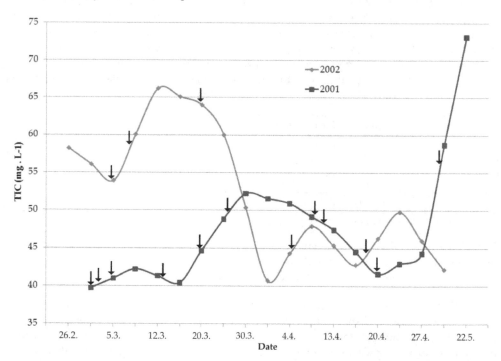

Fig. 1. Variation of total inorganic carbon (TIC) during application of pig slurry into the pond. Level of total doses 16 kg . m^{-2} in the year 2001 and 2002. (Arrows indicate term of pig slurry application)

Negative balance of carbon dioxide in ponds, evoked by plants assimilation at high content of nutrients in water, leads to high values of pH that could be the cause of fish gill necrosis. When using the pig slurry for ponds treatment, some bacteria and other microorganisms are also required to commence decomposition of the slurry and they have positive influence on zooplankton development, especially cladocerans. Bacteria that come into the water environment together with pig slurry are also a direct food source of zooplankton.

2. Fertilizer management

The amount of applied pig slurry depends on natural production of the pond, altitude, depth, control of the water inflow, fish stock etc. Due to the fish stock safety it is better to apply a slurry before vegetation season when a lower water temperature reduces the risk of oxygen deficit (Olah et al. 1986). Abnormal phytoplankton development is reduced in cold weather us well, which is associated with low pH and limited effect of toxic ammonia nitrogen on aquatic organisms. To maintain the stability of the pond ecosystem is not suitable to use a slurry in case of abnormal growth of submerged vegetation or water bloom, when water transparency is less than 40 cm (measured by Secchi disc), in case of abundance of basic nutrients or zooplankton overgrowth. For stability of the physical – chemical parameters it is necessary to properly organize an areal distribution, dose and interval of slurry application (Wohlfarth and Schroeder, 1979).

Knud-Hansen and Batterson (1994) tested four fertilization frequencies (daily, twice per week, weekly, and once every two weeks). The study showed that fertilization frequency had a strong linear relationship to net primary production. Zhu et al. (1990) showed that applying pig slurry daily produced higher yields of fish than five or seven day application. Garg and Bhatnagar (2000) also stated that fertilization frequency influenced fish yield. Variation of manuring frequency often depended on convenience, labour saving, the need to dispose the waste and availability of fertilizer, rather than increasing of the productivity of pond ecosystem. Under the experimental conditions it appears that the frequency of fertilizers application to maintain optimum primary production, should be every eight to ten days (Kestemont, 1995; Kumar et al. 2004). Planning of the fertilization frequencies should predict the time frame between each treatment which has got the potential of production enhancing and also reducing costs. There is a factor to be considered when fertilizing with manure: possibility of de-oxygenation of the water, which is quite likely when large quantities of manure are added at a time.

Fertilizers batching at doses 0.5-2.0 kg.m^{-2} of ponds water increased live weight gain of fish about 30-450 kg.ha^{-1} without supplementary feeding of fish (Hartman et al., 1973; Dhawan and Kaur, 2002). The effect of organic fertilizers from animal breeding farm is higher in ponds with polycultural fish stock where is the highest weight-gain per fish (Buck et al., 1978; Dhawan and Kaur, 2002; Zoccarato et al., 1995). Woynarovich (1976) reported pig slurry conversion of 3–5 % into fish body mass. In the case of polycultural fish stock is ratio higher than in the case of carp monoculture (Prinsloo and Schoonbee, 1984). Zhu et al. (1990) mentioned demand of 8.3 kg dry weight of pig slurry to weight-gain of 1 kg fish flesh. The maximal increases in fish production from fertilizers is especially in the tropics, on average, 4 kg fresh pig slurry can produce 1 kg of fish (Biro, 1995). Variously high doses of pig slurry were tried in ponds experiments. Pig slurry doses around 15-16 kg.m^{-2} were acceptable but they should be applied mainly before vegetation season, when a water temperature is lower (Kopp et al. 2008). Ponds have had concentrate polycultural fish stock during the experiment and fish were intensively fed by cereals. The increased production of the pond supported by pig slurry was 270–630 kg.ha-1. In these experiments the fish production in the high fertilized ponds appears the same values that obtained other researchers.

3. Autotrophs and heterotrophs

A successful fertilization programme in ponds should develop adequate amounts of food organisms for fish. Organic fertilizer (e.g. pig slurry) has been used to stimulate the development of heterotrophs (bacteria), autotrophs (algae) and other food organisms (zooplankton) to increase fish production in ponds (Schroeder, 1978). Bacteria and algae are important food organism for the herbivorous zooplankton that is consumed by many fish species (Wylie and Currie, 1991). In additional to the bulk of photosynthetic production passing directly to higher trophic levels, in some cases more than 50% of primary production is derived from a microbial loop (Hepher, 1992; Qin et al. 1995; Biro, 1995). Bacterivorous flagellates are a key link in the microbial loop to transfer organic carbon from bacteria to zooplankton and fish (Sanders and Porter, 1990). An increasing amount of evidence shows that bacteria together with algae provide an efficient energy pathway from low trophic levels to zooplankton. In natural waters, bacterial production can be enhanced by increasing organic matter loading, but the excessive application of organic matter into fish ponds can reduce dissolved oxygen and can cause fish kills (Qin and Culver, 1992). It is crucial to keep the amounts of organic fertilizer within limits such that adequate amount of heterotrophic organisms are produced yet without oxygen depletion occurring.

The reduction of algal biomass through grazing of zooplankton may indirectly regulate the amount of substrate available for bacterial growth. Bacterial and algal abundances were negatively correlated in fertilized ponds. This may have resulted from the die-off of algae that released dissolved organic compounds into the water, which in turn stimulated bacterial growth (Qin et al. 1995). In natural lakes, zooplankton populations crash after the decline of algae. In ponds, the zooplankton population might be sustained after algal decline by adding organic fertilizer to promote bacterial growth. Thus a direct input of pig slurry could increase bacterial productivity, which in turn could serve as a valuable food source for zooplankton.

When organic fertilizers are applied to ponds the first link of the food chain is heavily influenced, principally phytoplankton (Dhawan and Toor, 1989). High doses of pig slurry supported typically groups of phytoplankton favour higher amount of organic matter. Dominant algal genera often are *Cryptomonas*, *Chroomonas*, *Monodus* and *Euglena*. Generally also are occurred small genera of centric diatoms (*Stephanodiscus*, *Cyclotella*) and chlorococcal green algae (*Desmodesmus*, *Scenedesmus*, *Chlorella*, *Monoraphidium*). Cyanobacteria usually are not dominant groups in fertilizing ponds. High nitrogen to phosphorus ratios non favour dominance by cyanobacteria in pond phytoplankton (Smith, 1983). Typically genus *Microcystis*, *Aphanizomenon* and *Dolichospermum* in many fish-ponds and eutrophic lakes, cyanobacteria constitute the greater part of the summer phytoplankton biomass, causing regular water blooms and massive fish mortalities by depletion of oxygen after the bloom collapsed are not common in fertilizing ponds. Most often are occurred genus of cyanobacteria *Pseudanabaena, Planktothrix* and *Aphanocapsa* (Kopp and Sukop, 2003; Qui et al. 1995; Terziyski et al. 2007). Higher fish stock, sufficiency of carbon dioxide, suitable ratio of nutrients (C, N, P) in ponds after applications of pig slurry handicapped cyanobacteria forming water blooms. Cyanobacteria are of poor food value to zooplankton, their large size making them inaccessible to the filter-feeding entomostraca. Even the substances produced by many species of cyanobacteria are toxic to aquatic plants and animals (Sevrin-Reyssac and Pletikosic, 1990).

In ponds treated with slurry they have positive influence on zooplankton development. Though, the results of the experiments indicate that the fish stock with its predaceous pressure exerts a more significant influence on zooplankton than application of the slurry. Nevertheless, the initial development of large species of cladocerans was in ponds fertilized by slurry the more intensive to compare with control ponds (Sukop, 1980). When comparing different treatments, zooplankton was significantly higher in ponds manured with pig slurry. Higher zooplankton density as a result of manuring has also been reported in ponds receiving pig dung (Govind et al. 1978; Dhawan and Kaur, 2002). Dominant of zooplankton genera often are Copepoda (*Cyclops, Thermocyclops* and *Acanthocyclops*), nauplius and copepodits stages, small and middle genus of cladocerans (*Bosmina, Chydorus, Moina* and *Daphnia*) and Rotatoria (*Brachionus, Keratella* and *Asplanchna*) (Kopp and Sukop, 2003; Terziyski et al. 2007). Development biomass of planktonic communities in pond after pig slurry application is demonstrated in Figures 2-4.

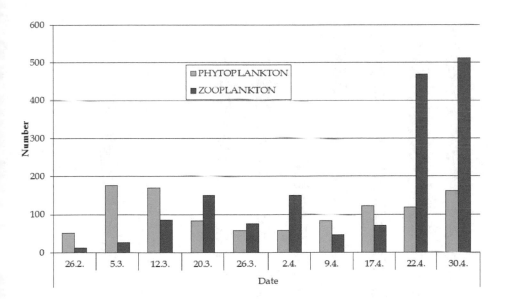

Fig. 2. Quantity of phytoplankton (individuals 10^3 . mL^{-1}) and zooplankton (individuals . L^{-1}) in pond during applications of pig slurry in total doses 16 kg . m^{-2}. (Application began at 5.3.)

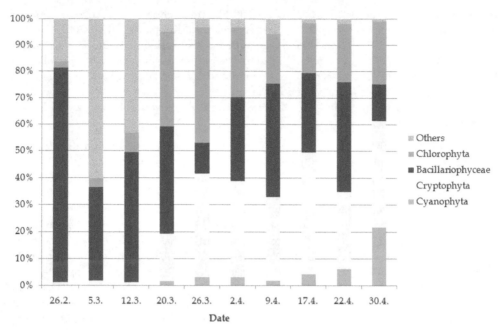

Fig. 3. Quantity of main groups of phytoplankton in pond during applications of pig slurry in total doses 16 kg . m-2. (Application began at 5.3.)

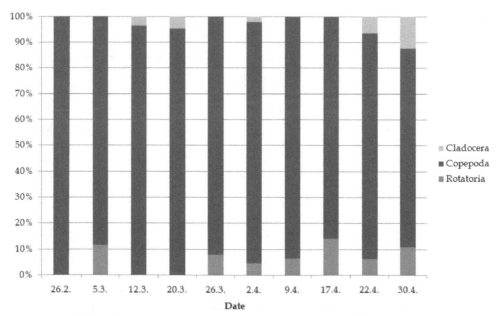

Fig. 4. Quantity of main groups of zooplankton in pond during applications of pig slurry in total doses 16 kg . m-2. (Application began at 5.3.)

4. Organic matter

Application of pig slurry to the water environment is bringing a huge amount of organic matter. Compare to other organic fertiliser the pig slurry has got lower content of organic substances, usually not exceed 10% in total (Hennig and Poppe, 1975). Majority of organic substances in pig slurry is biodegradable and relatively quickly mineralized by microorganisms. During the higher doses of manure to the fish ponds is important to control oxygen content which is intensively consumed due the decomposition of organic matter (Kopp et al. 2008). Application of organic fertiliser into the pond is immediately reflected in the increase of biological oxygen demand (BOD) and chemical oxygen demand (COD). As shows Figures 5 and 6, values change in wide range depending on dosage and elapsed time since last application. Important role for degradation of organic matter plays mainly water temperature and amount of microorganisms that decompose present organic matter. Significant reduction of COD and BOD observed during pig slurry application (Fig. 5 and 6) were mainly caused by changes of water temperature and the rate of degradation of organic substances. In water unloaded organic substances is value of BOD under 8 mg . L-1 and COD under 35 mg . L-1, due to manuring could be these values doubled (Zalud, 2008). Just one month after last application of pig slurry into the pond are decreasing values of organic load to a level comparable with ponds without fertilization (Kopp et al. 2008).

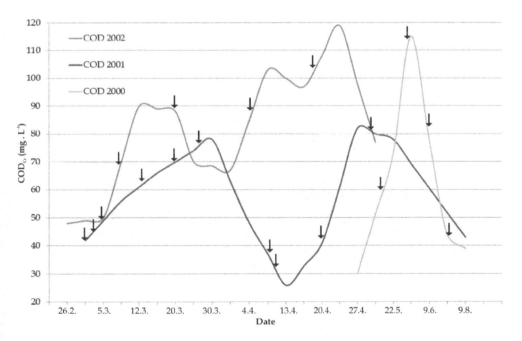

Fig. 5. Variation values of chemical oxygen demand (COD$_{Cr}$) during application of pig slurry into the ponds. Level of total doses 16 kg . m-2 (2001, 2002) and 0.10 kg . m-2 (2000). (Arrows indicate term of application of pig slurry)

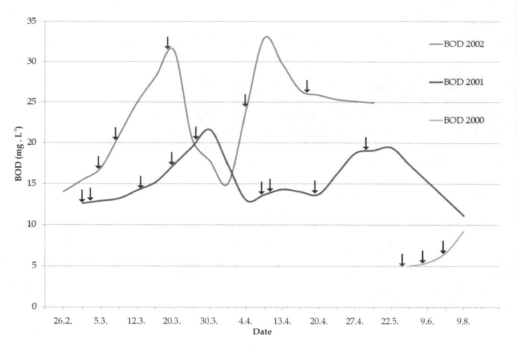

Fig. 6. Variation values of biological oxygen demand (BOD₅) during application of pig slurry into the ponds. Level of total doses 16 kg . m⁻² (2001, 2002) and 0.10 kg . m⁻² (2000). (Arrows indicate term of application of pig slurry)

5. Chlorophyll-a

Chlorophyll content in surface water is an important indicator of trophic level and one of the most often used indicators of biomass of primary producers. Concentration of chlorophyll in clear water usually does not exceed 10 μg . L⁻¹. In fish ponds and water reservoirs are values during the vegetation season in the tens to hundreds μg . L⁻¹. In case of hypertrophic water area with massive cyanobacteria water bloom could be chlorophyll content over thousands μg . L⁻¹ (Zalud, 2008).

The average amount of phytoplankton in ponds, expressed as chlorophyll-a concentration, has increased since 1960s from 30-35 μg . L⁻¹ to 140-150 μg . L⁻¹ till then 1990s. The impact of organic fertilization on increasing values of chlorophyll-a is well documented (Potužák et al., 2007). The application of organic fertilizer at regular intervals caused regular fluctuations of chlorophyll-a in the ponds.

After the application of pig slurry is increasing phytoplankton biomass as well as content of chlorophyll-a. The subsequent increase of zooplankton is lowering phytoplankton biomass by predation and also cause decreasing of chlorophyll–a. Fluctuations in value of chlorophyll–a during the pig slurry application is shows in Figure 7. Despite bigger doses of pig slurry are values of chlorophyll–a not so high. High stock of carp fish is increasing inorganic water turbidity by the feed pressure on the ponds bottom, decreasing transparency and limiting abnormal phytoplankton development due to inadequate lighting conditions. Lower water temperature and predation of zooplankton is limiting phytoplankton as well. (Kopp et al. 2008).

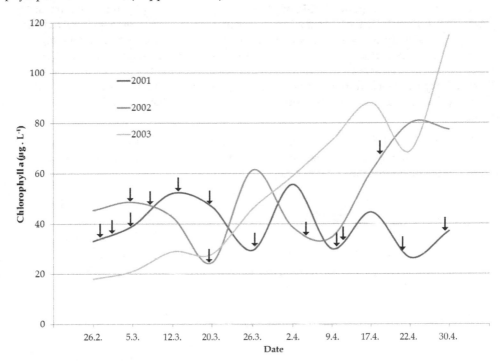

Fig. 7. Variation values of chlorophyll-a during the application of pig slurry into the pond. Level of total doses 16 kg . m-2 (2001, 2002) and without application of pig slurry (2003). (Application began at 28.2. in the year 2001 and 5.3. in the year 2002 respectively)

6. Dissolved oxygen

Dissolved oxygen concentrations in the ponds are most affected by phytoplankton biomass. Greater oxygen production and consumption occurs in ponds with higher phytoplankton biomass. The use of pig slurry increased the biological oxygen demand in ponds and could result in periods of low dissolved oxygen levels. The aerobic decomposition of organic matter by bacteria is an important drain of oxygen supplies in ponds. When added to ponds organic fertilizers exert an oxygen demand and an excessive application may result in depletion of dissolved oxygen (Schroeder, 1974, 1975). Qin et al., (1995) observed that dissolved oxygen in fish ponds with organic fertilizer was lower than in those without organic fertilizer, and dissolved oxygen in enclosures decreased with the increase of organic fertilizer loading.

Oxygen consumption in pond is regulated by biological and sediment oxygen demand and fish. In intensive fish aquaculture, BOD reached 0.29 mg O_2 . L^{-1} . h^{-1} in Israel (Schroeder, 1975), and 0.14 mg O_2 . L^{-1} . h^{-1} in fertilized ponds in USA (Boyd 1973). Qin et al., (1995) described that the BOD value varied with organic matter loadings and temperature. The BOD of fertilized ponds water varied from 0.19 to 0.34 mg O_2 . L^{-1} . h^{-1}. Similar results described Kopp et al. (2008) from Czech Republic, when the BOD value varied from 0.12 to 0.34 mg O_2 . L^{-1} . h^{-1} in pond during application high doses of pig slurry (Figure 8). It is certain that excessive organic fertilizer inputs can be a major factor depleting oxygen in ponds, especially during the clear-water phase when algae are less abundant (Qin and Culver, 1992). Analogous situation can be turn up during application of organic fertilizers, when higher development of zooplankton reduced the quantity of phytoplankton and dissolved oxygen concentrations declined. At high temperatures a heavy organic fertilizer application should also be avoided, otherwise a fish kill is likely to occur due to oxygen depletion.

On the second hand Dhawan and Kaur, (2002) reported that pig slurry even at higher dose (36 t . ha^{-1} . yr^{-1}) had no adverse effect on the dissolved oxygen content, which is an important parameter for the survival and growth of fish.

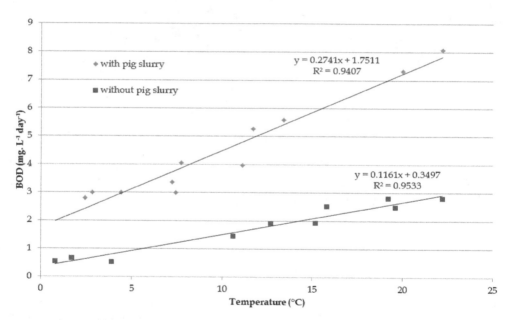

Fig. 8. Regression of temperature with biological oxygen demand (BOD). Pond water without pig slurry was used from pond a year before application and pond water with pig slurry was used during application of 16 kg . m^{-2} of pig slurry (February-May)

Sharma and Das, (1988) also reported that in a fish-pig pond, even higher organic loading by pig slurry did not deteriorate the oxygen content of water. This may be attributed to the higher plankton abundance in manure-loaded ponds, leading to higher dissolved oxygen level by photosynthetic activity especially during day. The values of dissolved oxygen varied during day and night in large interval and after application of pig slurry in during higher development of phytoplankton biomass can be exceed to 250 % saturation within day

(Behrends et al. 1980; Kopp et al., 2008). Figure 9 shows fluctuations in oxygen content during the application of pig slurry at low and high values. In both cases oxygen varied in wide range but has not fall below 50% of saturation, which is limit value ensure survival of carp fish. A wide range of values of oxygen during the application of pig slurry is determined by the intensity of photosynthesis, which is mainly influenced by the amount of available nutrients, water temperature and feed pressure of the zooplankton.

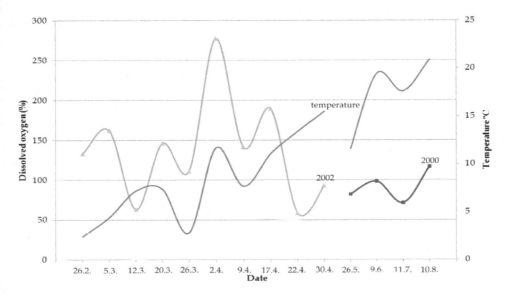

Fig. 9. Variation values of dissolved oxygen during application of pig slurry into the ponds. Level of total doses 16 kg . m^{-2} (2002) and 0.10 kg . m^{-2} (2000) (application began at 9.5. in the year 2000 and 5.3. in the year 2002 respectively)

7. pH

The pH value of water depends on the alkalinity and hardness of the water. Algal photosynthesis causes an increase in pH due to removal of H$^+$ ions. Algal assimilation of nitrate and its subsequent reduction within the algal cell to ammonium also increases the pH. On the second hand pH decreases with organic fertilization because bacterially generated carbon dioxide from manure decomposition. It was clear that the fertilization regimes were able to maintain adequate buffering capacity so that the pH fluctuations were within the desired limits for fish culture.

Properly chosen dosage and interval of slurry application should keep pH in optimum range for fish farming. To avoid pH increasing (over 8.5) due to intensive photosynthesis of phytoplankton what can happen especially in pond with a low alkalinity it is recommended to add calcium. Dose of calcium fertilizer is set up along of water alkalinity and depends on amount and timing of slurry application. In intensively fertilized ponds is pH usually under 9.0 (Kopp et al. 2008). Figure 10 shows pH fluctuations during the high dose application of pig slurry into intensively managed pond for two years. Significant changes in pH depending on phytoplankton development were apparent in 2001. First two years after slurry application is decreasing pH due to reduced nutrients supply and lower development of primary producers.

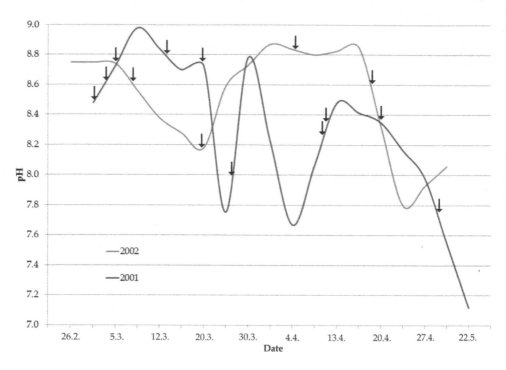

Fig. 10. Variation values of pH during application of pig slurry into the pond. Level of total doses 16 kg . m^{-2} (2001, 2002). (Arrows indicate term of pig slurry application)

8. Transparency

Application of fertilizer at regular intervals affects the transparency of the water. The transparency was significantly worsened within several weeks in the ponds after pig slurry application. Organic particles from manure had a significant influence in water turbidity. Minimum transparency was measured in ponds with abnormal phytoplankton biomass and higher fish stock of cyprinids, which reducing transparency by feeding by pressure on the bottom. High abundance of autotrophs may cause higher development of zooplankton and better transparency. During applications of pig slurry was transparency usually in interval 20-40 cm (Kumar et al. 2004; Kopp et al. 2008). Figure 11 shows different transparency during the high dose application in 2001 and 2002 compare to water transparency in 2003 without pig slurry application. Higher transparency in 2003 was due to lower level of nutrients and lower phytoplankton development in cooler period of the year compare to years with pig slurry application.

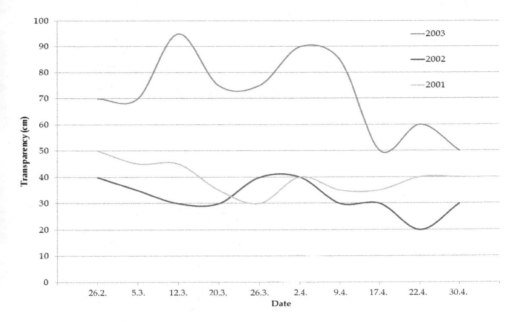

Fig. 11. Variation values of transparency during application of pig slurry into the pond. Level of total doses 16 kg . m⁻² (2001, 2002) and without application of pig slurry (2003). (Application began at 28.2. in the year 2001 and 5.3. in the year 2002 respectively)

9. Alkalinity

The alkalinity of the water is associated with the level of carbonates and bicarbonates content. The bicarbonate and carbon dioxide in the water has a buffering effect, preventing sudden changes in pH. Low alkalinity (under 1 mmol . L⁻¹) leads to poor buffering capacity and generally, a high fluctuation of pH. A high buffering capacity prevents fluctuation in the acid-base equilibrium caused by strong photosynthetic activity. Alkalinity increases with pig slurry application because bacterially generated carbon dioxide from manure decomposition dissolves calcium carbonate present in the pond sediments and slurry (Kumar et al. 2005). The gradual reduction in alkalinity occurred during the prime period of primary production. According to Boston et al. (1989), a decrease in pond water alkalinity can occur when algae remove bicarbonates. This may occur during periods of high photosynthetic activity which is promoted by application of pig slurry. Figure 12 shows fluctuations in alkalinity during the application of pig slurry into the fishpond. Alkalinity rises at the beginning of application (in accordance with literature) and after that change due to phytoplankton biomass development. Significant decrease of alkalinity during the pig slurry application in 2002 was caused by abnormal plankton development, high intensity of photosynthesis and depletion of carbonates. The alkalinity was quite stable or fluctuates during the year in small interval only in the fishponds that was not fertilizer.

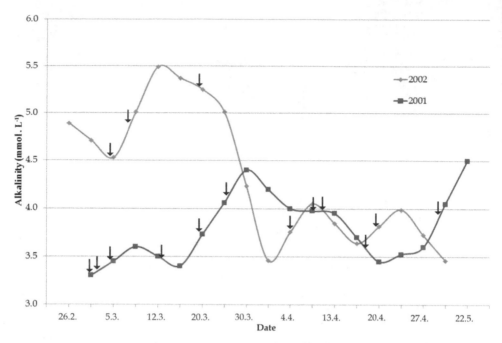

Fig. 12. Variation values of alkalinity during application of pig slurry into the ponds. Level of total doses 16 kg . m⁻² (2001, 2002). (Arrows indicate term of application pig slurry)

10. Nutrients

The nutrient content of animal manure may vary with time an d the quantity of nutrient released from the manures over time is a key factor in determining the fertilization regime. According to Muck and Steenhuis (1982), manure nutrient concentrations, and the percentage of P, N and C which become available for algal uptake, depend primarily on the animal´s diet, whether the manure is liquid or solid, and the age and storage conditions of the manure. The primary productivity pattern in the pond indicated a gradual built-up phase, a peak phase followed by a decline phase, this can be related to the nutrient release pattern. In pig slurry application, the primary productivity attained its peak 5-7 days after the application of slurry (Kumar et al. 2004).

Culver (1991) and Culver et al. (1993) introduced a fertilization regimen which attempts to optimize the algal composition by maintaining nitrogen and phosphorus at a specific ratio in ponds. They suggested that green algae could be maintained whereas cyanobacterial blooms could be suppressed in ponds by manipulating the N:P ratio (20:1) with fertilizer. Oin et al. (1995) showed that it is difficult to reach the N:P ratio recommended by Culver et al. (1993) by using only organic fertilizer, because most organic fertilizers have a low N and P content and inappropriately low N:P ratios.

Sediment is usually considered the major sink of orthophosphates in fishponds. The sediments absorb soluble forms of P from pond water until fully saturated. The rate of absorption decreases with an increase in partial saturation of sediments (Boyd and Musig, 1981). The P probably accumulates in the newly created sludge layer. Thus, the retained P does not appear to seep into the underground water, but remains in the pond. The P-containing sediment is the basis of the fertility of the pond. The higher P- input into the pond, the more P was retained (Knosche et al. 2000). Pig slurry has a low P content relatively, and application of pig slurry to pond has not negatively impact on higher value of P in outlet water (Kopp et al., 2008; Blažková et al., 1987). Balances of total P and N during the application of pig slurry into the ponds are demonstrated in the Figure 13 and 14.

Nitrogen is one of the most important agents of eutrophication, but it is not the most decisive factor in the eutrophication process of surface waters. Its balance is also influenced by the fixation of N from the air caused by cyanobacteria and by bacterial denitrification. The N balances in carp pond shows the similar picture as the P-retention, i.e. a clear increase in retention with N load (Olah et al. 1994). On the second hand Pechar (2000) reported that the rise in P and N in carp ponds during the 1990s is a result of the higher loading of organic fertilizers, particularly pig and cattle manure. Higher doses of pig slurry increase amount of nitrogen in outlet water from manure pond, especially ammonium for define time (Kopp et al. 2008).

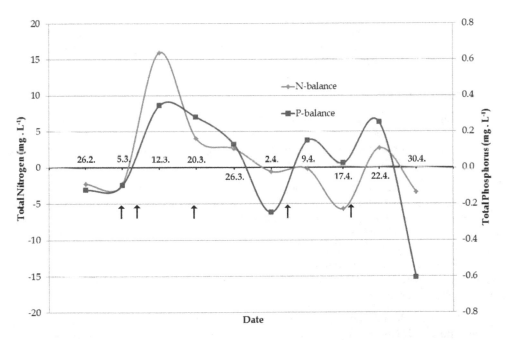

Fig. 13. Balance of total P and N (output by outflow minus input by inflow) during the application of pig slurry (total doses 16 kg . m^{-2}) into the pond (February-April), (Arrows indicate term of application pig slurry)

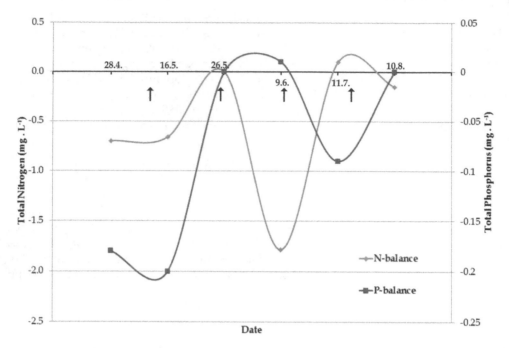

Fig. 14. Balance of total P and N (output by outflow minus input by inflow) during the application of pig slurry (total doses 0.10 kg . m⁻²) into the pond (May-July), (Arrows indicate term of application pig slurry)

11. Ammonia nitrogen

The critical water quality parameter during application of pig slurry is ammonia nitrogen, a principal excretory product of fish metabolism, which in the un-ionized form is highly toxic to fish. The oxidation of ammonia requires oxygen in the water, for this reason during the initial period of pig slurry application, low levels of nitrate were recorded.

The amount of toxic ammonia depends mainly on pH and water temperature. The maximum allowable (safe) concentration for carp is very low (0.05 mg . L^{-1} NH_3), LC_{50} (lethal concentration) for carp is 1.0 -1.5 mg . L^{-1} NH_3 (Svobodová et al. 1971; Russo and Thurson, 1991). During the application of higher dose of pig slurry has been level of ammonia nitrate increased for a short time which could endanger fish stock by not just acute toxic effects of ammonia but by autointoxication of fish body as well (Kopp et al. 2008). The dosage of pig slurry must follow hydrochemical conditions in pond and application is not possible in high pH or higher water temperature.

Figure 15 shows fluctuations of ammonia nitrogen during the high dose application of pig slurry. The graph clearly demonstrates the significant increase of ammonia in 2002, when were applied higher doses of slurry then a year ago. Despite high level of toxic ammonia in the water there was no fish kills or noticeable reduction of food intake but we can expect some negative impact of sublethal level of ammonia on fish organism. Arrilo et al. (1981)

observed biochemical changes in fish after 48 hour exposure to a low concentration of toxic ammonia (0.02-0.04 mg . L⁻¹). Higher value of toxic ammonia in combination with high pH is show by reduced weight, lower ability to survive and worse feed conversion (Máchová et al. 1983).

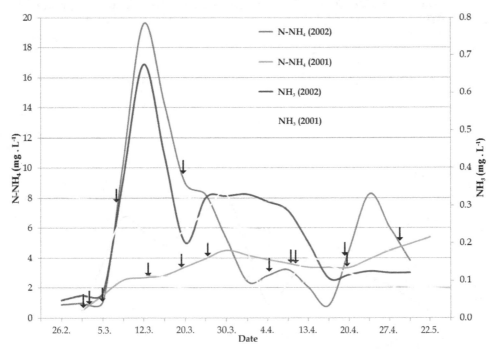

Fig. 15. Variation values of ammonia nitrogen during application of pig slurry into the ponds. Level of total doses 16 kg . m⁻² (2001, 2002). (Arrows indicate term of application pig slurry)

12. Nitrate and nitrite

The amount of nitrates which gets into the aquatic environment during the slurry application is relatively small compare to ammonia nitrate. Primary producers are using nitrates as sources of nitrogen therefore is negative balance between nitrates intake (water supply, organic fertilizer) and drain from the fishpond. High concentration of nitrates has not negative effect on fish stock but significantly contributes on water eutrophication. In Figure 16 is well show an increase of nitrogen after the start of slurry application and decrease after fertilization. Level of nitrate nitrogen in eutrophic pond often decreases under measurable value due to high assimilation of phytoplankton during the vegetation season (Kestemont, 1995).

Nitrite is an intermediate stage in the oxidation of ammonium to nitrate. Elevated nitrite concentration in water is a potential threat to freshwater fish since nitrite is actively taken up via the gills in competition with chloride and causing elevation of methaemoglobin levels (Russo and Thurson, 1991). Application of pig slurry to the water is bringing a lot of nitrate nitrogen. As show Figure 16, even during the high dose of slurry the value of nitrates is not dangerous for fish organism. Due to their chemical and biochemical instability nitrites in toxic environment are quickly transformed into nitrates and use by phytoplankton. Even in case of short term high value of nitrites due to single application of pig slurry fish are protected against the toxic effect of chloride ions which are a normal part of surface water (Jensen, 2003).

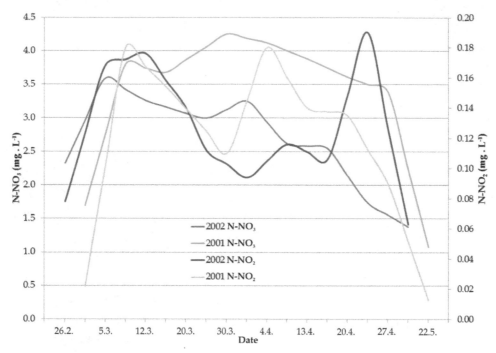

Fig. 16. Variation values of nitrate and nitrite during application of pig slurry into the pond. Level of total doses 16 kg . m⁻² (2001, 2002). (beginning application at 28.2. in the year 2001 and 5.3. in the year 2002 respectively)

13. Conclusion

Dhawan and Kaur, (2002) presented that pig dung, as pond manure which even at higher dose did not adversely affect water quality. Likewise Kajak and Rybak, (1980) advised, than even the heaviest load of nutrients from pig slurry which was added to water did not destroyed the functioning of the ecosystem until an "ecological catastrophe" resulted. A significantly higher plankton production was also recorded in this treatment. However, indiscriminate use of pig dung may deteriorate the water quality and hence decrease plankton production (Boyd and Doyle 1984). Preliminary determination of doses and composition of organic fertilizer to be introduced seems advisable. Another solution will might be able to allow enough time to recover the desired water quality after the introduction of organic fertilizer. The enormous quantities of organic fertilizers and the high fish stock densities in ponds resulted the presence of very small-sized group of zooplankton.

The phytoplankton blooms elevated pH levels and decreased the N:P ratio. Cyanobacteria, *Planktothrix agardhii* and *Limnothrix redekei*, typical of hypertrophic waters have become common (Pechar, 1995).

Pig slurry is used especially to supplement carbon into water. Due to change of environmental condition by human, eutrophication of fish ponds is increasing. Uncontrolled and unbalanced input of nutrients causes a disparity of basic biogenic elements N, P, which are in excess to carbon. High assimilation of plants makes carbon a limiting element for another production. This situation is common especially in ponds with intensive fish management (Sukop, 1980). It is necessary to implement the ameliorative intervention in colder period of the year considering higher hazard of variations of decisive hydrochemical parameters at higher water temperature. Unsuitable influence of high single doses of pig slurry on hydrochemical parameters is evident especially by higher values of toxic ammonia (Kopp et al. 2008). Pig slurry application has only short-term influence on water quality. Values describing that high organic pollution are noted immediately after the application. Adequate doses of pig slurry and acceptable form of its distribution to ponds does not cause permanent decline of water quality. Influence on values of physical and chemical parameters in water is not permanent (Sharma and Olah, 1986; Blažková et al., 1987).

From the water management point of view, ponds are not a burden to the environment, but generally improve the water quality downstream of the ponds. A claim for the reduction of production intensity in pond aquaculture cannot be justified from the water quality concerns. Carp ponds generally release better-quality water than these water bodies receive as inflow. Additionally, ponds act as water storage basin and improve the microclimate.

Public attention in recent years is focused on negative impact of agriculture on nature. High eutrophication of the water environment is decreasing biodiversity, causes abnormal water bloom development and fluctuations in physico-chemical parameters. So the application of organic fertiliser could be a problematic issue thought positive impact on fish production. Considering to the pressure of the human society on better environment the application of this kind of matter (i.e. pig slurry) into the water is strictly restricted in many countries.

14. Acknowledgment

This chapter was supported by the Research plan No. MSM6215648905 "Biological and technological aspects of sustainability of controlled ecosystems and their adaptability to climate change", which is financed by the Ministry of Education, Youth and Sports of the Czech Republic.

15. References

Arillo, A., Margiocco, C., Melodia, F., Mensi, P., Schenone, G. (1981). Biochemical aspects of water quality criteria: the case of ammonia pollution. Environment Technology Letters 2:285-292

Behrends, L.L., Maddox, J.J., Madewell, E.C., Pile, S.R. (1980). Comparison of two methods of using liquid swine manure as an organic fertilizer in the production of filter-feeding fish. *Aquaculture* 20:147-153

Biro, P. (1995). Management of pond ecosystems and trophic webs. *Aquaculture* 129:373-386

Blažková, D., Kočková, E., Žáková, E. (1987). Effect of slurry application and changes of water quality in ponds. In: *Intensification of fish production and water quality*. Velké Meziříčí 8.–9. December 1987, s. 56–61 (in Czech)

Boston, H.L., Adams, M.S., Madsen, J.D. (1989). Photosynthetic strategies and productivity in aquatic system. *Aquatic Botany* 34:27-57

Boyd, C.E. (1973). The chemical oxygen demand of waters and biological material from ponds. *Transactions of the American Fisheries Society* 102:606-611

Boyd, C.E., Musig Y. (1981). Orthophosphate uptake by phytoplankton and sediment. *Aquaculture* 22:165-173

Boyd, C.A., Doyle, K.M. (1984). The timing of inorganic fertilization of sunfish ponds. *Aquaculture* 37:169-177

Buck, D.H., Baur, R. J., Rose, C.R. (1978). Utilization of swine manure in a polyculture of Asian and North American fishes. *Transactions of the American Fisheries Society* 107, 1: 216–222

Culver, D.A. (1991). Effects of the N:P ratio in fertilizer for fish hatchery ponds. *Verhandlungen Internationale Vereinigung fur Theoretische und Angewandte Limnologie* 24: 1503-1507

Culver, D.A., Madon, S.P., Qin, J. (1993). Percid pond production techniques: timing, enrichment, and stocking density manipulation. *Journal of Applied Aquaculture* 2:9-31

Dhawan, A., Toor, H.S. (1989). Impact of organic manure and supplementary diet on plankton production and fish growth and fecundity of an Indian major carp, *Cirrhina mrigala* (Ham.) in fishponds. *Biological Wastes* 29: 289-298

Dhawan, A., Kaur, S. (2002). Effect of pig dung on water quality and polyculture of carp species during winter and summer. *Aquaculture International* 10 (4): 297–307

Garg, S.A., Bhatnagar, A. (2000). Effect of fertilization frequency on pond productivity and fish biomass in still eater ponds stocked with *Cirrhinus mrigala* (Ham.). *Aquaculture Research* 31:409-414

Govind, B.V., Raja, G., Singh, G.S. (1978). Studies on the efficacy of organic manures as fish feed producers. *Journal of Inland Fisheries Society* 10:101-106

Hartman, P., Lavický, K., Červinka, S., Pokorný, J., Komárková, J., Reichard, S. (1973). The use of pig slurry for ponds fertilization. – *Report of State fishing* Č. Budějovice, 24 p. (in Czech)

Hepher, B. (1962). Primary production in fishponds and its application to fertilization experiments. *Limnology and Oceanography* 7: 131-137

Hennig, A., Poppe, S. (1975). Abprodukte tierischer Herkunft als Futtermittel. VEB Deutcher Landwirtschaftsverlag, DDR 104, Berlin, 302 pp. (in German)

Jensen, F.B. (2003). Nitrite disrupts multiple physiological functions in aquatic animals. *Comparative Biochemistry and Physiology* – Part A, 135: 9-24

Kajak, Z., Rybak, J. (1980). The effect of pig manure and mineral fertilization on a eutrophic lake ecosystem. *Development of hydrobiology* 2: 337-346

Kestemont, P. (1995). Different systems of carp production and their impacts on the environment. *Aquaculture* 129:347-372

Knösche, R., Schreckenbach, K., Pfeifer, M., Weissenbach, H. (2000). Balances of phosphorus and nitrogen in carp ponds. *Fisheries Management and Ecology* 7:15-22

Knud-Hansen, C.F., Battevrson, T.R. (1994). Effect of fertilization frequency on the production of Nile tilapia (*Oreochromis niloticus*). *Aquaculture* 123: 271-280

Kopp, R., Sukop, I. (2003). The development of plankton communities at applications of pig liquid manure on Jarohněvický pond. *Acta Facultatis Ecologiae*, 10, Suppl. 1: 271-273 (in Czech)

Kopp, R., Ziková, A., Mareš, J., Vítek, T. (2008). Variations of chemical parameters in hypertrophic pond within pig slurry application. *Acta Universitatis agriculturae et silviculturae Mendelianae Brunensis* 2: 95-99

Kumar, S.M., Luu, T.L., Ha, V.M., Dieu, Q.N. (2004). The Nutrient Profile in Organic Fertilizers: Biological Response to Nitrogen and Phosphorus Management in Tanks. *Journal of Applied Aquaculture* 16:45-60

Kumar, S.M., Binh, T.T., Luu, T.L., Clarke, M.S. (2005). Evaluation of Fosh Production Using Organic and Inorganic Fertilizer: Application to Grass Carp Polyculture. *Journal of Applied Aquaculture* 17:19-34

Máchová, J., Peňáz, M., Kouřil, J., Hamáčková, J., Macháček, J., Groch, L. (1983). The Effects of Different pH Values and Increased Ammonia Concentrations on the Growth and Ontogenetic Development of Carp Fry. *Bulletin VÚRH Vodňany* 3: 3-14 (in Czech)

Muck, R.E., Steenhuis, S.T. (1982). Nitrogen losses from manure storage. *Agricultural Wastes* 4:41-54

Olah, J., Sinha, V.R.P., Ayyappan, S., Purushothaman, C.S., Radheyshyam, S. (1986). Primary production and fish yields in fish-ponds under different management practices. *Aquaculture* 58: 111-122

Olah, J., Pekar, F., Szabo, P. (1994). Nitrogen cycling and retention in fish-cum-livestock ponds. *Journal of Applied Aquaculture* 10: 342-348

Pechar, L. (1995). Long-term changes in fish pond management as an uplanned ecosystem experiment: importance of zooplankton structure, nutrients and light for species composition of cyanobacterial blooms. *Water Science and Technology* 32:187-196

Pechar, L. (2000). Impacts of long-term changes in fishery management on the trophic level water quality in Czech fish ponds. *Fisheries Management and Ecology* 7:23-31

Potužák, J., Hůda, J., Pechar, L. (2007). Changes in fish production effectivity in eutrophic fishponds-impact of zooplankton structure. *Aquaculture International* 15:201-210

Prinsloo, J.F., Schoonbee, H.J. (1984). Observation on fish growth in polyculture during late summer and autumn in fish ponds at the Umtata Dam Fish Research Centre, Transkei. Part I: The use of pig manure with and without pelleted fish. *Journal Water SA* 10: 15-23

Qin, J., Culver, D.A., Yu, N. (1995). Effect of organic fertilizer on heterotrophs and autotrophs: implications for water quality management. *Aquaculture Research* 26:911-920

Qin, J., Culver, D.A. (1992). The survival and growth of larval walleye, *Stizostedion vitreum*, and trophic dynamics in fertilized ponds. *Aquaculture* 108:257-276

Russo, R.C., Thurson, R.V. (1991). Toxicity of ammonia, nitrite and nitrate to fishes. Pages 58-59 in D.E. Burne and J.R. Tomasso, editors. Aquaculture and water quality. Word Aquaculture society, Baton Rouge, Lousiana

Sanders, R.W., Porter, K.G. (1990). Bacterivorous flagellates as food resources for the freshwater crustacean zooplankter *Daphnia ambigua*. *Limnology and Oceanography* 35:188-191

Schroeder, G.L. (1974). Use of fluid cowshed manure in fish ponds. *Bamidgeh* 26: 84-96

Schroeder, G.L. (1975). Night-time material balance for oxygen in fish ponds receiving organic wastes. *Bamidgeh* 27: 65-74

Schroeder, G.L. (1978). Autotrophic and heterotrophic production of microorganisms in intensively-manured fish ponds, and related fish yields. *Aquaculture* 14: 303-325

Sevrin-Reyssac, J., Pletikosic, M. (1990). Cyanobacteria in fish ponds. *Aquaculture* 86:1-20

Sharma, B.K., Olah, J. (1986). Integrated fish-pig farming in India and Hungary. *Aquaculture* 54:135-139

Sharma, B.K., Das, M.K. (1988). Studies on integrated fish-livestock carp farming system. *Fishing Chimes* 7:15-27

Smith, V.H. (1983). Low nitrogen to phosphorus ratios favour dominance by blue-green algae in lake phytoplankton. *Science* 221: 669-671

Sukop, I. (1980). Effect of Application of Poultry Slurry and Cererite on the Zooplankton Development in Fingerling Ponds. *Czech Journal of Animal Science* 25, 11: 847–855

Svobodová, Z., Groch, L. (1971). Possibilities of the Diagnosis of Ammonia-Intoxication of Fish. *Bulletin VÚRH Vodňany* 1: 9-18 (in Czech)

Terziyski, D., Grozev, G., Kalchev, R., Stoeva, A. (2007). Effect of organic fertilizer on plankton primary productivity in fish ponds. *Aquaculture International* 15:181-190

Wohlfarth, G.W., Schroeder, G.L. (1979). Use of manure in fish farming – a review. *Agricultural Wastes* 1: 279-299

Woynarovich, E. (1976). The possibility of combining animal husbandry with fish farming, with special reference to duck and pig production. – FAO *Techn. Conf. Aquacult.*, Kyoto, Japan 1976, R6: 11 s.

Wyllie, J.L., Currie, D.J. (1991). The relative importance of bacteria and algae as food sources for crustacean zooplankton. *Limnology and Oceanography* 36: 708-728

Zalud, Z. (Eds.) (2008). Biological and technological aspects of sustainability of controlled ecosystems and their adaptability to climate change- indicators of ecosystem services. *Folia Universitatis Agriculturae et Silviculturae Mendelianae Brunensis* 4: 176 pp. (in Czech)

Zhu, Y., Yang, Y., Wan, J., Hua, D., Mathias, J. A. (1990). The effect of manure application rate and frequency upon fish yield in integrated fish farm ponds. *Aquaculture* 91 (3–4): 233–251

Zoccarato, I., Benatti, G., Calvi, S.L., Bianchini, M.L. (1995). Use of pig manure as fertilizer with and without supplement feed in pond carp production in Northern Italy. – *Aquaculture* 129, 1–4: 387–390

6

An Ecotoxicological Approach to Evaluate the Environmental Quality of Inland Waters

M. Guida[4], O. De Castro[1], S. Leva[4], L. Copia[4],
G.D'Acunzi[3], F. Landi[3], M. Inglese[2] and R.A. Nastro[5]
[1]Department of Biological Sciences-University "FedericoII" of Naples,
[2]Laboratory of Environmental and Food Analyses Studies "Inglese & co. ltd",
[3]Salerno District Administration Bureau,
[4]Department of Structural and Functional Biology,University "Federico II"of Naples,
[5]Department of Sciences for the Environment, University Parthenope of Naples,
Italy

1. Introduction

Water is traditionally considered a renewable resource as the quantity theoretically available depends essentially on meteoric water contributions. Modern theories consider river ecosystem as a central element of the environment but, actually, water ecosystems are more and more polluted because of agricultural and industrial activities. In Italy, where there is a widespread urbanization, most part of river ecosystems are exposed to a severe risk of damage, with consequent loss of biodiversity. The institution of national and regional parks makes possible the preservation of significant natural value areas. In 1974, the UNESCO established a "World Network of Biosphere Reserves" (WNBR) aimed to preserve areas, where there is a close relation between man and nature, through environmental preservation and sustainable development (UNESCO,2010). Nevertheless, a constant monitoring along the whole watercourse (from spring to mouth) is needed as, even in a protected area, human activities could cause, directly or indirectly, damages to valuable ecosystems.

1.1 The ecotoxicology

In recent years, ecotoxicology started a new approach to the environmental analysis as it covers:

- chemistry, i.e. the fate of chemicals in the environment
- environmental toxicology, dealing with the evaluation of toxic effects of a pollutant at different levels of biological integration
- ecology, which provides indications on regulation of both structure and function of ecosystems and, at the same time, on interactions between biotic and abiotic components (De Castro et al., 2007).

Ecotoxicology is based on the use of bio-indicators belonging to different levels of an ecosystem trophyc chain. Mathematical models are also used to foresee the environmental fate of chemicals and their effects on exposed organisms (man included) and ecosystems. An organism can be considered a" bio-indicator" when, in presence of pollutants, shows detectable variations from its natural state. Well defined responses to different concentrations of pollutants are also needed. Moreover, a good bio-indicator should be sensitive to pollutants and have a wide distribution in the investigated environment, low mobility, long life cycle and genetic uniformity. Actually, *Daphnia magna, Lepidium sativum, Cucumis sativus, Sorghum saccharatum, Pseudokirchneriella subcapitata, Vibrio fischeri* are widely used to evaluate water and soil quality as well as the toxicity of chemicals, wastes, pharmaceutical products which have to be processed in a wastewater plant or directly dumped in the environment. The main symptoms or endpoints of an ecotoxicological test could be:

- change in community structure;
- morphological changes;
- change in vitality;
- damage to genes.

Actually, a contemporary utilization of different bio-indicators in order to evaluate the ecotoxicity of a matrix (wastewater, contaminated soil, pharmaceutical by-products, etc.) let the researchers able to gain useful data about the possible toxic effects on an ecosystem (Guida et al., 2006). In the next lines, informations about biology of some test-organisms adopted in ecotoxicological analyses and about the main indices and parameters used in the evaluation of a river ecosystem quality are resumed.

1.1.1 *Daphnia magna*

Daphnia magna is a freshwater crustacean belonging to the class Brachiopoda, order Cladocera, phylum Arthropoda. It has a small size (not more than 5 mm in length) with an oval body compressed laterally. It is dorsally characterized by a welded bivalve structure (called "carapace"), which encloses the entire animal except for the head. Its body has a single compound eye, sessile, strongly pigmented; a small ocellus and two pairs of antennae. One pair of antennae is greedy and very developed, having essentially a swimming function. The transparency of carapace allows an observer to notice some internal organs: heart, located dorsally in the post-cephalic region, middle intestine (clearly visible when it contains food) and ovaries, that are located laterally. At a temperature of 20°C, the life cycle of *D. magna* is 60-100 days. Under favourable environmental conditions, the population is exclusively composed of female animals, with a parthenogenetic reproduction. This species is very sensitive to many pollutants able to cause variations to aquatic ecosystem, since it is considered the "perfect" bioindicator and test-organisms in ecotoxicological essays (Guida et al.,2004).

1.1.2 *Vibrio fisheri*

Vibrio fischeri is a rod-shaped Proteobacterium, gram negative, characterized by polar flagella. It is widespread in marine environments, living as symbiotic of various marine animals, such as the bobtail squid. *V. fischeri* is most often found as a symbiont of *Euprymna scolopes*, a small shallow water squid found on the shores of Hawaii.

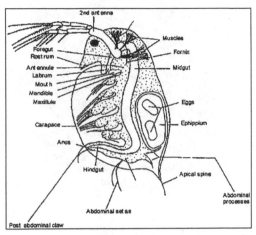

Fig. 1. A female individual of *Daphnia magna*: internal and external anatomy. (FAO, 1996)

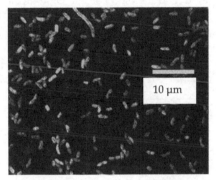

Fig. 2. The bioluminescence of *Vibrio fisheri* (1000X). Image taken by E. Nelson and L. Sycuro (http://microbewiki.kenyon.edu; last accessed: September 2011).

The bioluminescent bacterium *Vibrio fischeri* and juveniles of *Euprymna scolopes* specifically recognize and respond to each other during the formation of a persistent colonization within the host's nascent light-emitting organ. The biolumiscence depends on metabolic activity of bacteria so, damaged microrganisms less in biolumiscence. This feature has been used to study toxicity in marine samples.

1.1.3 *Pseudokirchneriella subcapitata*

Pseudokirchneriella subcapitata (previously called *Selenastrum capricornutum*) is a single-celled freshwater algae, belonging to the *Chlorococcales* family. It is a representative species of oligotrophic and eutrophic aquatic system. The cells, sickle-shaped, have a volume of 40-60 μm^3, size of 6-7 μm, and its life cycle is very short. This alga shows a good level of sensitivity to toxicants and it is commonly used to perform multispecies tests (Chen C.Y. et al., 2007).

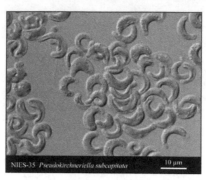

Fig. 3. Image at light microscope of *P.subcapitata* (1000x). (http://www.shigen.nig.ac.jp; last accessed: September 2011)

1.1.4 Phytotoxicity essays

The toxicity of a pollutant on a soil can be assessed by specific plants which, in natural ecosystems, can be considered good bioindicators. Ecotoxicological bioassays on plants consists generally in simple methods. It is possible to assess different endpoints for each essay such as seeds survival, germination rate, growth speed in the light and in the dark. Different plants could be used in ecotoxicological tests depending on the investigated matrix. Nevertheless, plants very commonly used in such essays in Italy and in some other Countries are *Lepidium sativum* and *Cucumis sativus* (Youn-Joo An; 2004).

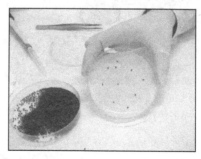

Fig. 4. Phytotoxicity test on *L. sativum*.

1.2 The evaluation of watercourse ecological quality: Extended Biotic Index (EBI) and Fluvial Functional Index (FFI)

If ecotoxicology allows scientists to gain informations about the potential toxicity of a single pollutant or a mixture of chemicals (like industrial wastewater) through the implementations of laboratorial tests, a complete study of a water ecosystems requires the collection of data directly on site. So, analyses of microfauna (small invertebrates colonizing water ecosystems) and ichtyofauna, the measure of parameters like water Dissolved Oxygen (DO), Turbidity, pH, Conductivity, Temperature, besides of quality of riverside flora and evaluation of watercourse erosion are needed to evaluate the environmental quality of a water ecosystem like a river. The collection and characterization of microfauna is at the basis of the Extended Biotic index (EBI), whose value is used to classify the quality of freshwater courses in Italy (Ghetti P., 1997). Further biotic indices based on the quali-quantitative analysis of invertebrates community are also used with the same aim (Kalyoncu H. and Zeybek M., 2011). Once calculated the EBI value (ranging from 0 to 12), a class of environmental quality is attributed to each stream or river or part of it. There are 5 classes available, corresponding to a well defined colour (Table 1). Colours are used to report the classification of environmental quality on a map of the watercourse. Like EBI, Fluvial Functioning Index (FFI) values correspond to well defined quality classes (Negri P.et al., 2011). Recently, National Environmental Protection Agency of Italy suggested FFI for a correct evaluation of the health of river ecosystems (APAT, 2007). It must be underlined that FFI categories are as recommended by Water Framework Directive 2000/60 (Negri P. et al., 2011).

Classes	E.B.I. (Values)	Outcome	Colour
Class I	10-11-12...	Watercourse not significantly polluted or not polluted at all	Blue
Class II	8-9	Watercourse moderately polluted or altered	Green
Class III	6-7	Watercourse polluted or altered	Yellow
Class IV	4-5	Watercourse highly polluted or altered	Brown
Class V	1-2-3	Watercourse extraordinarily polluted of altered	Red

Table 1. Classes of environmental quality corresponding to EBI values.

The above mentioned indices are based on a simply concept: a watercourse is a dynamic system, formed by different habitats which continuously follow each other from the source down to the mouth and which interconnect with the surrounding terrestrial ecosystems. It is important to notice how, along the course of a river, the environmental conditions (like morphological, hydrodynamic, physical and chemical parameters) change and, with respect to these, the biological populations vary too. That's why the state of a river ecosystem must be evaluated along the whole course in pre-determinate stations. The study of the whole territory crossed by the river and the identification of possible critical situation is of a great importance as it could affect heavily the results of a monitoring activity. The hydro-geologic conformation of the river is also important as it could affect the accessibility to sampling stations. So, an adequate preliminary study, a good experience in the choice of sampling sites, besides of a suitable equipment, are necessary.

1.3 Other indices

According to EU Regulation 2000/60/CE published on 10/23/2000 (the Water Framework Directive), the quality of a watercourse has to be evaluated through both Ecological and Chemical States indices. From the elaboration of these indices, the Ecological State of Watercourse (ESW Index) could be calculated. Another index widely used is the Evaluation of Macro-descriptors Pollution Level (EMPL) which is calculated from the following parameters: Chemical Oxygen Demand (COD), Biochemical Oxygen Demand at 5 days (BOD$_5$), NO$_3$-, NH$_4$+,Total Phosphorous, *Escherichia coli*, Oxygen saturation rate.

1.4 Evaluation of freshwaters microbial quality

Water quality evaluation can't exempt from a microbiological characterization of water itself. The presence of some bacterial species like *Escherichia coli* gives important informations about the grade of a watercourse pollution (Edberg S.C. et al., 2000). Different microbial parameters were considered in the present survey as more parameters led more informations about the organic pollution of a river ecosystem. *E. coli* and fecal coliforms are indicators of recent fecal pollution. Streptococci and Clostridia, instead, are abundant in case of a past fecal pollution. *Staphylococcus aureus, Pseudomonas aeruginosa, Aeromonas hydrophila* are potential pathogens of both fishes and men (Pathak S.P. et al., 2008; Health Canada, 2011; Fazli M., 2009).

1.5 Phytoplankton

Phytoplankton includes different algal species belonging to different taxa (*Bacillariophyceae, Dinophyceae, Chrysophyceae, Cryptophyceae, Dictyochophyceae, Prymnesiophyceae, Raphydophyceae and Euglenophyceae, Prasinophyceae and Chlorophyceae*). The analysis of the phytoplanctonic component of a water ecosystem, gives important informations about the amount of nutrients (mainly nitrates and phosphates), the presence of toxics and some chemical-physical factors as water temperature and turbidity. Moreover, phytoplankton populations and corporations oscillate widely even in short time or in small space according to the environmental conditions (as nutrients, dissolved oxygen, turbidity, temperature, etc.). That's why phytoplankton is considered, since long time, an important indicator of the trophic level of an aquatic ecosystem (freshwater and marine waters) (Greene J.C et al., 1975; Mahoney J.B., 1983).

2. Aim of the study

Ecotoxicological tests are usually carried out to assess the potential toxicity of wastes or chemicals before their introduction in the environment. In the course of two years, we tried to apply an ecotoxicological approach to evaluate the environmental quality of Tanagro and Bussento rivers flowing through a WNBR area (UNESCO, 2010), in Southern Italy. We used ecotoxicological tests findings to identify possible critic situations (due to a possible chemical pollution) along both river courses. But, since a river is a complex system, a polyphasic approach in a correct environmental quality evaluation was needed. In fact, the real challenge researchers have to face consists in consider all possible elements of river ecosystem (water quality, fauna and flora composition, erosion phenomena, etc.) in order to have a realistic picture of the ecological state of a watercourse. So, besides of ecotoxicological essays, we collected data about chemical, microbiological and physical characteristics of water while ecological quality of watercourses was evaluated by the use of EBI. Moreover, further important data were gained through the interpolation of different parameters, calculating, this way, ecological indices like Evaluation of Macro-descriptors Pollution Level (EMPL) and Ecological State of Watercourses (ESW). At last, we tried to obtain a complete picture of the river environmental quality. The dependability of ecotoxicological tests for river quater quality evaluation was assessed.

3. Material and methods

Standardized procedures were applied in sampling procedures and water analyses. A complete list of all parameters measured and indices calculated is reported in Table 2. The choice of sampling sites was carried out according to EU Regulation 2000/60/CE recommendations, taking in account the introduction of waste waters through spillways . On the whole, 14 sampling points (stations) were chosen for each river. Collected data were stored in a database and georefentiated by an ArcGis software (data not shown). Frequency of sampling and analytic procedures are reported in Table 3.

Chemical parameters	Microbiological parameters	Ecological and Ecotoxicological parameters
Temperature (°C)	Aerobic Colony Count at 22°C	Daphnia magna acute toxicity essay
Dissolved Oxygen (DO)	Aerobic Colony Count at 37°C	Pseudokirchneriella subcapitata acute toxicity test
Ph	Total Coliforms	Phytotoxicity test
Specific Conductivity (SC)	Fecal Coliforms	Inihibition test on algeae
NH_4^+	Escherichia coli	Fitoplankton
NO_2^-	Streptococci	Extended Biotic Index (EBI)
NO_3^-	Clostridia	Fluvial Functional Index (FFI)
PO_4^-	Staphylococci	
BOD_5	Stafilococcus aureus	
COD	Pseudomonas aeruginosa	
	Aeromonas hydrophila	

Table 2. Complete list of analyses carried out on each water sample analyzed.

Biological Parameters	Sampling frequency	Chemical-physical parameters	Sampling frequency
Ecotoxicological essays	5 times/year	Water temperature	5 times/year
Microbiological analyses	5 times/year	Dissolved Oxygen	5 times/year
Fitoplancton	3 times/year	Salinity	5 times/year
EBI	2 times/year	pH	5 times/year
FFI	1 time/year	Nutrients (NO_3^-,NO_2^-)	5Times/year

Table 3. Frequency of sampling for each parameter in each station.

3.1 Chemical analyses

The chemical analyses were carried out according to Standard Methods (2006).

3.2 Microbiological analyses

All microbiological analyses were carried out according to ISO methods.

3.3 Phytoplankton

The EN 15204:2006 and the EN ISO 5667-1 and EN ISO 5667-3 were applied during this survey.

3.4 *Daphnia magna* – Acute bioassay

Daphnia magna acute toxicity test were carried out according to ISO 6341:1999. According to this method, ten newborns no older than 24 hours must be exposed to a sample. The newborns of D. *magna* are transferred in each container filled with 50 ml of sample: this operation must be carried out paying attention to don't damage the daphnides. Moreover, they must not be fed during all test procedure. After 24 hours, the number of immobile crustaceans (or of the ones showing at least a change in their usual way of swimming) is calculated. If the essay is carried out considering different sample concentrations, it is possible to calculate the EC_{50} , which gives, for a well defined toxic, the concentration value inhibiting the 50% of test organisms.

3.5 *Vibrio fisheri* – Acute bioassay

APAT IRSA-CNR 2003 n. 8030 is the method chosen to carry out acute toxic bioassays with V. *fischeri*. The method gives an evaluation of acute toxicity of freshwater and marine samples through the evaluation of V. *fischeri*, strain NRRL-B-11177 bioluminescence inhibition. Luminescence can be measured after 5, 15 and 30 minutes of exposition to a sample by the use of a luminometer. Sample toxicity is measured as EC50, which represents the sample concentration in correspondence of which there is a decrease of 50% of the light emitted by bacteria. V. *fisheri* toxicity test demonstrated a good correlation with tests carried out on other aquatic organisms like D. *magna*, *Artemia salina*, *Chlorella* sp., *Tetrahymena pyriformis* (Kaiser K.L., 1998). Its reliability for the evaluation of soil and colourful samples had been also demonstrated (Lappalainen J. et al., 2001).

3.6 Chronic bioassay

ISO 8692:2004 is the method applied for *Pseudokirchneriella subcapitata* bioassay. The bioassay analyzes the toxic effect of a sample by measuring the inhibition of algal growth. Selected cultures are exposed during their exponential phase of growth, at well defined concentrations of a sample for 72 hours. After the exposition to the sample, algae density is measured reading absorbance of the culture at 663 nm.

3.7 Phytotoxicity

Phytotoxicity test assesses the potential toxicity of a sample measuring the inhibition of germination and /or root elongation of seeds under controlled conditions. Negative controls are prepared. Seeds of two dicotyledons (*L. sativum* and *C. sativus*) and a monocotyledon (*S. saccharatum*) are exposed to a water sample and incubated in the dark at 25 ± 2 °C for 72 hours. Then, germinated seeds are counted and root length measured using a calibre. The effect on both germination and radical elongation is expressed as percentage germination index (GI%) (UNICHIM 1651:2003). Such tests are widely used to assess the ecotoxicological effects of soils and waters contaminated with organic molecules and/or heavy metals (An Y. et al., 2004).

4. Results

4.1 Ecotoxicological results

Ecotoxicological tests didn't show any important inhibitory effect resulting in a good quality of both river waters. So, *D. magna* didn't suffer any significant toxic effect: the percentage of immobility didn't ever overcome, on the average, the 20% of individuals. Sampling sites number 7 and 13 of Tanagro river showed a modest effect on *P.subcapitata* (an increase in cell reproduction rate of about 10% on average). As to *L. sativum*, *C. sativus* and *S. saccharatum* no toxic effects were detected in both river samples. In some cases, especially for *L. sativum* and *S. saccharatum* some kind of bio-stimulation was observed (Fig.5 and 6). The results of phytotoxicity tests didn't show any important inhibition of seeds germination or

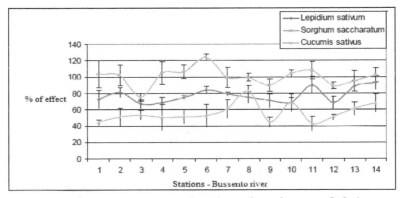

Fig. 5. Bussento river phytotoxicity tests on *L. sativum*, *S. saccharatum*, *C. Sativus*.

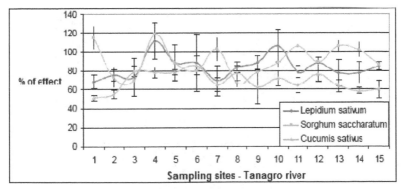

Fig. 6. Tanagro river phytoxicity tests findings. *C.sativus* showed some inhibition effects at station 1 and 14 while *S.saccharatum* was stimulated mainly in stations 1 and 4.

root elongation on the most part of the samples, even if *C. sativus* showed a less growth rate than the other two plants. This could be explained on the base of different physiology of the species used in the test. All samples were tested also on *V. fisheri* and they never showed any inhibitory effect on bacteria. It is important to notice that, in no case and for no test organism, it was possible to calculate EC50 value because of the really low samples toxicity.

4.2 Chemical parameters and EMPL

In all stations, COD values were, on the average, always less than 20 mgO_2/mL, while BOD_5 never overcome 10 mgO_2/mL in Bussento river (Fig.7). As to Tanagro river, an increase of COD and BOD_5 was detected at station number 3, whose values overcame 20 mgO_2/L for COD and 10 mgO_2/L for BOD_5 (Fig.8). Moreover, chemical analyses showed some significant variations of nitrogenous compounds (as NO_3^-), not only among the stations but

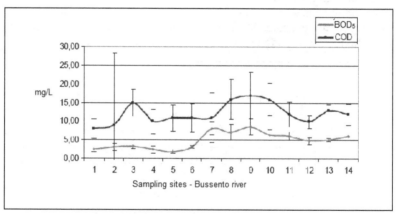

Fig. 7. BOD_5 and COD values in Bussento river. Average values and standard error are reported.

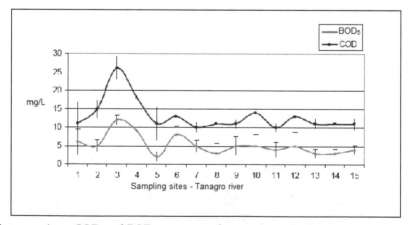

Fig. 8. Tanagro river: COD and BOD_5 average values and standard errors.

even for a single sampling site in the course of the time, causing an increase in variability (Fig. 9 and 10). So, the greatest variations were detected in correspondence of the stations number 6 and 7 of Bussento river and 3 and 4 for Tanagro ones. Nevertheless, as *E. coli* concentration never overcame the 5000 CFU/100mL limit, the increase of nitrates could be due to agricultural rather than to wastewater intakes. Both rivers, in fact, flow through a not highly urbanized land characterized by an agriculture-based economy.

If NO_2^- values were low in all samples, NH_4^+ concentration showed some different values between the two watercourses, as shown in figures 9 and 10. On the average, NH_4^+ was higher in Tanagro rather than in Bussento water. In any case, NH_4^+ concentration decreased from station number 10 till the river month. A similar tendency was detected for PO_4^{2-}, suggesting an agricultural origin of both nutrients.

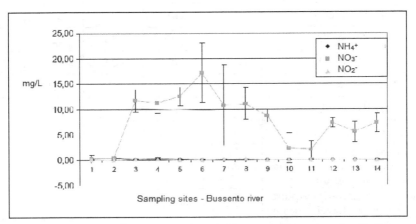

Fig. 9. Nitrogenous compounds concentrations in Bussento water samples. In the graphic, average values and standard errors are reported.

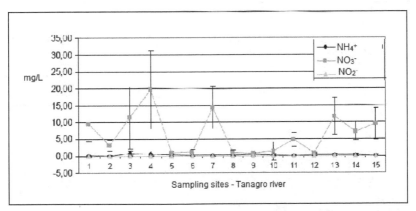

Fig. 10. Nitrogenous compounds concentrations (average ± standard error) along Tanagro river course.

No water acidification was detected in both rivers as pH values showed little variation and, in any case, they ranged between 7.5-8.5 values. The results of chemical analyses, together with *E. coli* concentrations, carried out on both rivers were compared to values shown in table 4, in order to calculate the EMPL index for both watercourses. From collected data, all stations were classified as belonging to the 2nd and 3rd levels, showing a moderate pollution. Furthermore, Tanagro river showed a better water quality than Bussento one as 73% of collected samples belonged to level 2 versus a 53% of Bussento ones. On the whole, 66% of all samples belonged to level 2.

Parameters	Level 1	Level 2	Level 3	Level 4	Level 5
100-DO (% sat.)	≤|10|	≤|20|	≤|30|	≤|50|	≥|50|
BOD$_5$ (O$_2$/mg/L)	<2.5	≤4	≤8	≤15	>15
COD (O$_2$/mg/L)	<5	≤10	≤15	≤25	>25
NH$_4^+$ (N mg/L)	<0.03	≤0.1	≤0.5	≤1.5	>1.5
NO$_3^-$ (N mg/L)	<0.30	≤1.5	≤5	≤10	>10
Total-P (P mg/L)	<0.07	≤0.15	≤0.30	≤0.6	>0.6
Escherichia coli (CFU/mL)	<100	≤1000	≤5000	≤20000	>20000
Score referable to each parameter (75° percentile of the sampling period)	80	40	20	10	5
Evaluation of Macro-descriptors Pollution Level	480-560	240-475	120-235	60-115	<60

Table 4. Reference values of EMPL index.

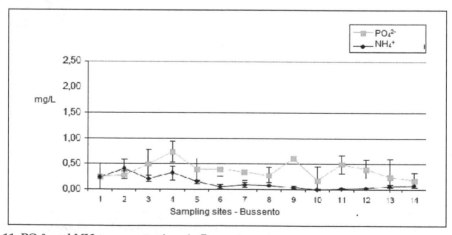

Fig. 11. PO$_4^{2-}$ and NH$_4^+$ concentrations in Bussento waters.

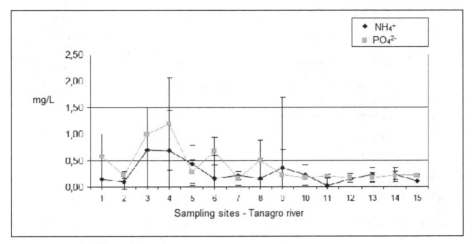

Fig. 12. PO_4^{2-} and NH_4^+ values along Tanagro river stations.

4.3 Extended biotic index

The analysis of macro-invertebrates population were carried out two times a year (on summer and on winter) in correspondence of the same stations used to take samples for chemical-physical and microbiological analyses. Temperature, DO, pH and water conductivity were measured contemporaneously as they affect macro-invertebrates distribution heavily and a correct EBI evaluation can not exempt from a determination of the above mentioned parameters. Even if it is possible to notice a similar trend in the EBI variation along both watercourses (from spring to mouth), the second sampling campaign (on summer) showed lower values than the first one (carried out on winter). Apart from few sites showing a real deterioration of river ecosystem (especially in Tanagro river), EBI average values didn't change significantly during the sampling activity. Most part of the collected samples were classified as belonging to the second class (corresponding to moderate pollution) or to the third one (altered ecosystem).

4.4 The Ecological State of Watercourses (ESW)

The EBI values together with EMBL ones were used to gain another index, the Ecologic State of Watercourses (ESW) (Table 5). Our results showed, on the whole, a good or sufficient ecological quality. Only 36% of Bussento river sampling sites reached the 2nd ESW class of quality, suggesting a certain vulnerability of the freshwater ecosystem itself

As the remaining sites resulted just in a 3rd class. Tanagro river was characterized by a better environmental state as 46% of stations were classified as belonging to a 2nd class. Nevertheless, our results shed light on a critical environmental state regarding the station 3 of Tanagro river (4th class of quality). Even if most part of ESW values were determined by both EBI and EMBL values, it is interesting to notice how, in some cases, ESW values were affected in most part by the EBI values rather than EMBL ones as reported in Table 6.

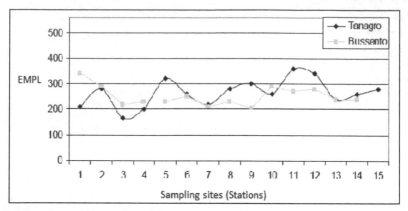

Fig. 13. EMPL values of Tanagro and Bussento river. There are not great variations in value.

The biological index not only is complementary to chemical characterization of freshwaters but it could even indicate, in advance, possible environmental criticisms.

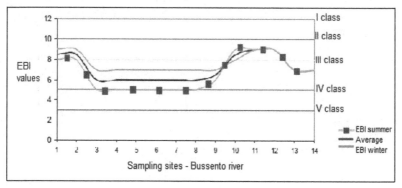

Fig. 14. EBI values of Bussento river stations. The values range from the 2nd to the 4 class of quality.

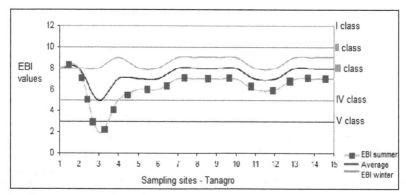

Fig. 15. Tanagro river EBI values in winter and in summer. In station n°3 the lowest value corresponding to a 5th class.

ESW	Class I	Class II	Class III	Class IV	Class V
EBI	≥10	8-9	6-7	4-5	1,2,3
EMBL	480-560	240-475	120-235	60-115	<60
Overcome	High	Good	Sufficient	Poor	Very Poor
Conventional colour	Blu	Green	Yellow	Orange	Poor

Table 5. ESW class of quality and the correspondent EBI and EMBL values are reported.

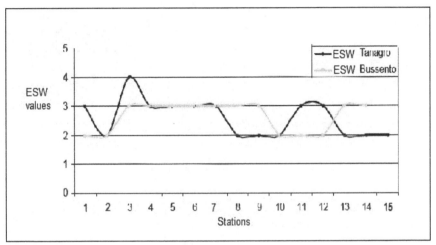

Fig. 16. ESW values: environmental quality reached the worst value just in station number 3 of Tanagro river.

	EMBL+EBI	EMBL	EBI
Tanagro	54%	13%	33%
Bussento	79%	21%	0%

Table 6. Percentage of samples whose ESW value was affected by EMBL+EBI or by just EMBL or EBI values.

FFI was applied in order to consider both hydro-morphological and biological factors in the evaluation of river courses environmental quality. Our findings, compared to some reference values, showed an environmental quality ranging from good-moderate to high (Table 7). The 57% of Tanagro river course fell into the high and good categories while Bussento river just 35%.

	High	Good	Good-moderate
Tanagro	7%	50%	43%
Bussento	21%	14%	65%

Table 7. FFI values for Tanagro and Bussento rivers.

4.5 Phytoplankton

Phytoplankton analyses didn't show any particular dystrophy, except for three stations of Bussento river, where high values of phytoplankton density were found.

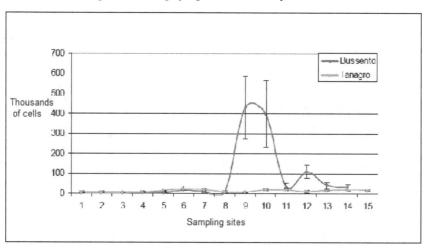

Fig. 17. Phytoplankton concentrations along the watercourses stations.

4.6 Microbiological results

Even if microbiological analyses are not strictly required to evaluate the environmental quality of a watercourse, the presence of some bacteria could be useful in order to recognize anthropogenic impacts due to wastewater intakes. As to *E. coli* concentration, Bussento water showed in all sites an amount ranging between 100 and 1000 CFU/100ml. On the other hand, most part of Tanagro samples (56%) showed a concentration ranging from 100 to 1000 CFU/100 ml, a 35% less than 100 CFU/100 ml and just a 9% of the stations was characterized by an amount overcoming 1000 CFU/100ml. These data were substantially confirmed by total and fecal coliforms, streptococci amounts. As to the other microbial parameters, no clostridia were found in both rivers water samples while *S.aureus* (typical of human and mammalian skin and mucosa) was seldom isolated. *Aeromonas hydrophyla* and *Pseudomonas aeruginosa*, instead, were widely present in water samples as they are part of environmental microflora. In Figures 18 and 19 the results concerning *A. hydrophyla* and *P. aeruginosa* are reported. From a comparison between both *A. hydrophyla*, *P.aeruginosa* and *E.coli* amounts along the rivers, there is no evidence of any correlation between such bacterial strains amounts.

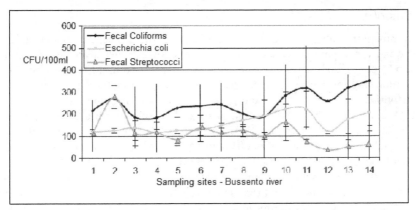

Fig. 18. Bussento river: microbiological values overcame the 100 CFU/100mL values except for streptococci in the last four stations.

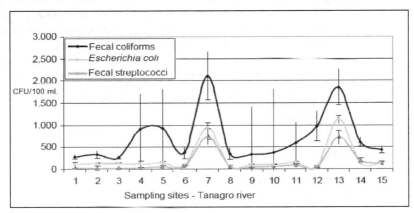

Fig. 19. *E.coli*, fecal coliforms and streptococci in Tanagro river waters. Stations 7 and 13 showed some high values.

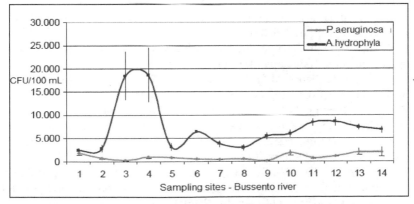

Fig. 20. *P. aeruginosa* and *A. hydrophila* concentrations in Bussento waters.

5. Conclusions

The environmental quality of both rivers resulted, on the whole, in a good or, at least, a sufficient quality, comparable, at least, to the upper course of Sele river flowing in the same area (Rizzo D. et al., 2009). These data are important as in a so highly urbanized district like the Campania one, most part of watercourses are heavily polluted and compromised by human activities. Sarno river, for example, is one of the most polluted river in Europe (Arienzo M. et al.,2000; De Pippo T. et al.,2006) and it flows in a high anthropized area. In this context of environmental degradation, the protection of high naturalistic value areas is a must and our data confirmed the efficacy of natural reserves and of a sustainable development policy. Under a strictly technical point of view, a polyphasic approach provides detailed informations about environmental quality of a river ecosystem. Nevertheless, it must be underlined that some experience in data analyses and competence in different fields are needed in order to give a right interpretation of findings yelded by so different analyses. In fact, it has to be noticed, for example, that phytotoxicity tests outcomes weren't fully overlapping with other ecotoxicological tests and of some difficult interpretation. While the *D.magna* and *P.subcapitata* tests gave no evidence of any significant toxic effect of water samples, *L.sativum, C.sativus, S.saccharatum* showed a modest inhibition of germination after 5 days (between 22% and 30% of seeds for each species didn't germinate) in correspondence to the station number 3 of Tanagro river, in accordance to the EBI, ESW, EMPL whose values were corresponding to a poor or sufficient environmental quality. It is clear that ecotoxicological tests are not enough to evaluate the environmental quality of a complex ecosystem like a river but they showed to be a useful tool in the evaluation of river environmental quality.

6. Acknowledgments

We thank Dr. Daniela Santafede for her contribution to the elaboration of chemical and microbiological data.

7. References

An Y. , Kim Y., Kwon T., Jeong S. (2004) Combined effect of copper, cadmium, and lead upon Cucumis sativus growth and bioaccumulation. Science of The Total Environment Volume 326, Issues 1-3, pp. 85-93 .

APAT (2007). Indice di Funzionalità Fluviale – Nuova versione del metodo revisionata ed aggiornata. http://info.apat.it/pubblicazioni

Arienzo M., Adamo P., Bianco M.R., Violante P. (2000). Impact of Land Use and Urban Runoff on the Contamination of the Sarno River Basin in Southwestern Italy. Water, Air, & Soil Pollution Volume 131, Numbers 1-4, 349-366

Autorità di Bacino Regionale Friuli Venezia Giulia (ABR.FVG). River Continuum Concept. http://www.abr.fvg.it/studies/ecologia-acque/water-ecology/river-continuum-concept Visited on January 2012.

Chen CY, Ko CW, Lee PI.(2007). Toxicity of substituted anilines to Pseudokirchneriella subcapitata and quantitative structure-activity relationship analysis for polar narcotics. Environ Toxicol Chem. Jun;26(6):1158-64.

De Castro O., Gianguzzi L., Colombo P., De Luca P., Marino G., Guida M.: (2007) Multivariate Analysis of Sites Using Water Invertebrates and Land use as Indicators of the Quality of Biotopes of Mediterranean Relic Plant (Petagnaea gussonei, Apiaceae) Environmental Bioindicators, 2, pp. 161 – 171.

De Pippo T., Donadio C., Guida M., Petrosino C. (2006). The case of Sarno River (Southern Italy): effects of geomorphology on the environmental impacts. Environ Sci Pollut Res Int. 2006 May;13(3):184-91.

Directive 2000/60/EC of the European Parliament and of the Council establishing a framework for the Community action in the field of water policy". Official Journal (OJ L 327). December 2000.

Edberg SC, Rice EW, Karlin RJ, Allen MJ. (2000). Escherichia coli: the best biological drinking water indicator for public health protection. Symp Ser Soc Appl Microbiol. 2000;(29):106S-116S.

FAO (1996). Manual on the Production and Use of Live Food for Aquaculture. FAO Fisheries Technical Paper 361, pp. 283-288.

Fazli M., Bjarnsholt T., Kirketerp-Møller K., Jørgensen B., Schou Andersen A., Krogfelt K.A., Givskov M., and Tolker-Nielsen T.(2009) Nonrandom Distribution of Pseudomonas aeruginosa and Staphylococcus aureus in Chronic Wounds. J Clin Microbiol.47(12): 4084–4089.

Ghetti PF (1997). Application Manual: Extended Biotic Index – Macroinvertebates in quality control of running water environments, in Italian [I Macroinvertebrati nel controllo della qualita di ambienti di acque correnti. Indise Biotico Esteso (I.B.E). Manuale di applicazione. Provincia Autonoma di Trento]. Trento Italy, 222 pp.

Greene J.C., Miller W.E., Shiroyama T. and Maloney T.E.(1975). Utilization of algal assays to assess the effects of municipal, industrial, and agricultural wastewater effluents upon phytoplankton production in the Snake River system. Water, Air, & Soil Pollution Volume 4, Numbers 3-4, 415-434.

Guida M., Mattei M., Melluso G., Pagano G., Meriç S.(2004) Daphnia magna and Selenastrum capricornutum in evaluating the toxicity of alum and polymer used in coagulation-flocculation" Fresenius Environmental Bulletin; vol 13 N° 11b 1244-7.

Guida M., Melluso G., Mattei M., Inglese M., Pagano G., Belgiorno V., Meriç S. (2006): A multi-battery lethal and sub-lethal toxicity investigation on organo-pesticides. 18-22 September Istanbul- Turkey. IWA DipCon 10th International Specialised Conference on Diffuse Pollution and Sustainable Basin Management.

Health Canada. Environmental and Workplace Health: Bacterial Waterborne Pathogens - Current and Emerging Organisms of Concern. www.hc-sc.gc.ca (visited on September 2011).

Kaiser K.L. (1998). Correlations of Vibrio fischeri bacteria test data with bioassay data for other organisms. Environ Health Perspect. 106 (Suppl 2): 583–591.

Kalyoncu H. and Zeybek M. (2011) An application of different biotic and diversity indices for assessing water quality: A case study in the Rivers Çukurca and Isparta (Turkey).African Journal of Agricultural Research Vol. 6(1), pp. 19-27.

Lappalainen J., Juvonen R., Nurmi J., Karp M. (2001) Automated color correction method for Vibrio fischeri toxicity test. Comparison of standard and kinetic assays Chemosphere Volume 45, Issues 4-5, Pages 635-64.

Mahoney J. B. (1983) Algal assay of relative abundance of phytoplankton nutrients in northeast United States coastal and shelf waters.

Negri P., Siligardi M.,Fuganti A.,Francescon M., Monauni C.,Pozzi S. The use of the Functioning Fluvial Index for river management.
evis.sggw.waw.pl/wethydro/contents/.../107-115_PaoloNegri_e.pdf (visited on September 2011).

Pathak S.P. , Bhattacherjee J.W., Kalra N., Chandra S. (2008). Seasonal distribution of Aeromonas hydrophila in river water and isolation from river fish. Journal of Appl. Microbiol. Vol. 65 Issue 4, pp. 347 – 352.

Rizzo D., Gentile A., Sica M, Caracciolo L.,Mautone M.,Rastrelli L.,Saviello G. Caratterizzazione Qualitativa e Tipizzazione del Fiume Sele Individuazione e definizione del Corpo Idrico di Riferimento Sito-Specifico. Atti del Convegno Funzionalità fluviale:strumento di gestione e pianificazione. Trento, 19-20 November 2009.

Youn-Joo An (2004) Soil ecotoxicity assessment using cadmium sensitive plants. Environmental Pollution Volume 127, Issue 1, Pages 21-26.

UNESCO – MAB Secretariat (2010). World Network of Biosphere Reserves.
http://www.unesco.org/mab/doc/brs/BRList2010.pdf visited on January 2012.

7

Emerging (Bio)Sensing Technology for Assessing and Monitoring Freshwater Contamination – Methods and Applications

Raquel B. Queirós[1,2], J.P. Noronha[3],
P.V.S. Marques[1] and M. Goreti F. Sales[2]
[1]INESC TEC (formerly INESC Porto) and Faculty of Sciences, University of Porto,
[2]BioMark Sensor Research and ISEP/IPP - School of Engineering,
Polytechnic Institute of Porto,
[3]REQUIMTE/CQFB and Faculty of Sciences and Technology of
University Nova de Lisboa,
Portugal

1. Introduction

Water is life and its preservation is not only a moral obligation but also a legal requirement. By 2030, global demands will exceed more than 40 % the existing resources and more than a third of the world's population will have to deal with water shortages (European Environmental Agency [EEA], 2010). Climate change effects on water resources will not help. Efforts are being made throughout Europe towards a reduced and efficient water use and prevention of any further deterioration of the quality of water (Eurostat, European Comission [EC], 2010). The Water Framework Directive (EC, 2000) lays down provisions for monitoring, assessing and classifying water quality. Supporting this, the Drinking Water sets standards for 48 microbiological and chemical parameters that must be monitored and tested regularly (EC, 1998). The Bathing Water Directive also sets concentration limits for microbiological pollutants in inland and coastal bathing waters (EC, 2006), addressing risks from algae and cyanobacteria contamination and faecal contamination, requiring immediate action, including the provision of information to the public, to prevent exposure. With these directives, among others, the European Union [EU] expects to offer its citizens, by 2015, fresh and coastal waters of good quality.

Freshwater quality is generally monitored with regard to chemical and microbiological parameters. Most regulated chemical parameters are monitored by more or less expensive but reliable techniques and some of which allow on-site analysis. The ease of access to such analytical procedures is reflected on the large amount of chemical data given by Water Information System for Europe [WISE] (EEA, 2011). Microbial parametric data are, on the contrary, more difficult to obtain. Microorganisms may be defined as those organisms that are not readily visible to the naked eye, requiring a microscope for detailed observation. These have a size range (maximum linear dimension) up to 200 µm, and vary from viruses, through bacteria and archea, to micro-algae, fungi and protozoa (Sigee, 2005).

There are many different microorganisms that can pose serious risks to the environment and public health (Figure 1). In general, waterborne pathogens cause 10–20 million deaths and 200 million non-fatal infections each year (Leonard et al., 2003). They contribute to harmful effects either by a direct or indirect way: through the direct contact with the pathogen (when they are ingested by susceptible men or other living organisms) or by their metabolites like the toxic products excreted to water bodies. Some of these contaminants of microbiological origin are subject to limiting values in waters.

Legal requirements indicate cultured-based methods to monitor cell activity or highly sophisticated and expensive instrumental-based methods for metabolite detection/quantification. In general, these methods are cumbersome and take too long to produce the desired response; within that time the contamination can move/spread out, while users of recreational waters and possible consumers are at risk of contracting serious infections. Regardless of their sensitivity and selectivity these methods also are unsuitable for routine and on-site applications (An & Carmichael, 1994; Health Protection Agency, 2005; International Organization for Standardization [ISO], 2005).

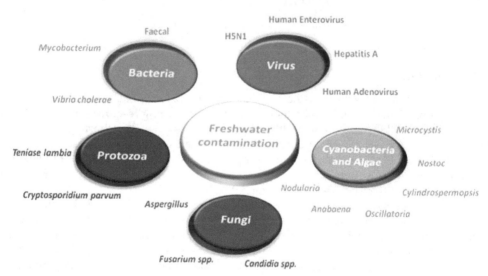

Fig. 1. Microorganisms contaminating freshwater.

Novel technologies for assessing and monitoring microbial water contamination are now becoming available. Biosensors are here of main interest (Figure 2). They have a capture-probe (or biorecognition element) on a standard transduction surface allowing on-site inexpensive determination. The biorecognition element (enzymes, cofactors, cells, antibodies, nucleic acids, or structured polymer) interacts with the target analyte, producing a physical-chemical change that is converted by the transducer into a detectable signal. This signal can be measured optically, mechanically, magnetically, thermally or electrically (Nayak et al., 2009; Su et al., 2011; Turner & Piletsky, 2005; Vo-Dinh, 2007).

Optical or radiant transducers can be classified according to mode or scattering. Classification by mode includes absorption, emission or combination thereof (absorbance or transmission in Ultra-Violet [UV], Visible [Vis], or Infra-Red [IR] region of the spectrum, Attenuated Total Reflectance [ATR], Evanescent Field, Surface Plasmon Resonance [SPR], Luminescence, and Photoemission). Classification by scattering consists on phase change, polarization, absorption and opto-thermal effect (Raman, Ellipsometry, SPR and Photo-Acoustic effect). Mechanical transducers are often called frequency transducers and include Surface Acoustic Wave [SAW], piezoelectric oscillators, and Quartz Crystal Microbalance [QCM]. Magnetic transducers are Nuclear Magnetic Resonance [NMR] and mass spectrometry detection systems. Thermal transducers include calorimetric systems. Electrical transducers can be called electrochemical transducers when the electrical measure is evaluated in a solution. The classification of electric transducers by mode can include voltage (Potenciometric) or current transducers (Amperometric), current-voltage transducers (Voltammetric, Field Effect Transistors [FETs], Metal Oxide Semiconductor [MOS], charge transfer and resistance transducers (coulometric, chemiresistors, ion mobility, and mass spectrometry) and dielectricity transducers (Capacity systems) (Spichiger-Keller, 1998).

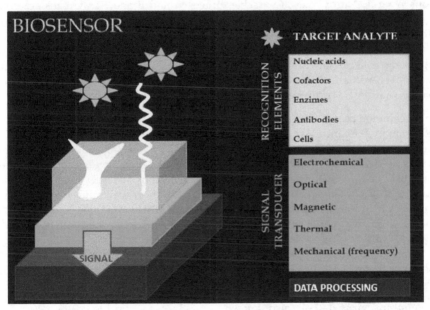

Fig. 2. General biosensor scheme.

Biosensors must meet essential requirements for a successful application. First, the output signal must be of relevance. The biosensor must be accurate and sensitive, providing zero "false-positives" and few "false negatives" and should be reproducible. Moreover the biosensor must be specific, discriminate between the target organism and other organisms. The response time of the biosensor must be inferior to the analytical reference methods, ideally providing a "real-time" response. Also, it must be physically robust and must be able to resist temperature, pH, and ionic strength changes as well as be insensitive to other environmental interferences. The assay should require minimal operator skills for routine detection. One of the most important criteria is the cost of either manufacture, running and life cost. Finally, the biosensing assay must be accurate against standard techniques (Leonard et al. 2003).

Biosensors have been applied by now to the most important forms of microbial contamination: virus, protozoa, bacteria and their toxins, harmful algae and fungi. This chapter will review emerging biosensing technology, methods and applications for assessing and monitoring microbial water contamination. It will focus the development of cost-effective on-site methods, based on biosensing devices that show the potential to complement both laboratory-based and field analytical methods. It will be structured according to biorecognition elements used in the biosensing system, and quote the microorganisms, metabolites or species that are a potential risk to humans; legal requirements; and emerging issues in water and infectious disease.

2. Virus biosensors

The most important waterborne viruses belong to Enteric viruses [EV]. Enterovirus, Hepatitis A virus, Noroviruses, Coxsackie viruses A and B and Rotaviruses are present in gastrointestinal tract of contaminated individuals and are eliminated through feces in larger quantities. Due to its incomplete removal from sewage (even after chlorination) EV are likely to be transmitted to humans from contaminated drinking water (Tavares, 2005). These pathogens can cause a variety of symptoms such as, non-bacterial gastroenteritis, infectious hepatitis, myocarditis and aseptic meningitis. EV can remain viable several months in water in adverse conditions (Tavares et al., 2005; Gilgen et al., 1997).

The most common assays for virus detection use antibody-based methods such as Enzyme-Linked ImmunoSorbant Assay [ELISA] (Bao et al. 2001), fluorescent antibody assays (Barenfanger et al., 2000) and serologic testing (O'Shea et al., 2005). PCR assays are frequently used as complement technique to improve the detection of low concentration of virus (Henkel et al., 1997), despite being a costly and laborious technique. Several authors (Brassard et al., 2005; Gilgen et al. 1997; Soule et al., 2000) presented Reverse Transcriptase – Polymerase Chain Reaction [RT-PCR] as sensitive and efficient method for detection of few EV. Since these diagnostic methods are generally unwieldy and often have limited sensitivity, a variety of new virus detection methods, including microcantilevers (Ilic et al., 2004), evanescent wave biosensors (Donaldson et al., 2004), immunosorbant electron microscopy (Zheng et al., 1999) and atomic force microscopy (Kuznetsov et al., 2005), have emerged to conquer these limitations.

Avian influenza virus [AIV], also known as H5N1, is considered by World Health Organization [WHO] an emerging issue and a potential risk of infection and recommends the monitoring of this virus in waters (WHO & WASH Inter Agency Group, 2007). The natural reservoirs of H5N1 virus are wild waterfowls; these birds may excret large quantities of the virus in their feces as well as in the saliva and nasal secretions. Infected migratory waterfowl may then enter water environments where the birds gather and transmit the low pathogenicity virus to other migratory waterfowl and to domestic birds. During replication, the virus can mutate and highly pathogenic avian influenza [HPAI] strains may occur. The main concern with respect to the HPAI H5N1 virus is that it may change into a form that is highly infectious for humans and that spreads easily from person to person (WHO, 2007).

AIV can persist for extended periods of time. H5N1 has been isolated from unconcentrated waters from lakes in Canada, United States and Hong Kong. There are no quantitative data available on levels of H5N1 virus in lake water where waterfowl gather, although its detection in unconcentrated waters and in small sample volumes suggest relatively high levels. Besides direct deposition of faeces into lake waters by migratory waterfowl, it has been suggested that faecal waste from duck and chicken farms may spread to bodies of water via wind, surface runoff or possibly enter groundwater through disposal and composting of waste on poultry farms (WHO, 2007).

Some authors reported a few new biosensors approaches for the rapid detection of H5N1 virus using impedance measurement of immuno-reaction coupled with red blood cells amplification (Lassiter et al., 2009; Lum & Li, 2010), interferometric immunobiosensor (Xu et al., 2007) and electrochemical-aptamer based biosensor (Liu et al., 2011). Lum & Li developed an impedimetric biosensor for the detection of AIV. The biosensor is composed by immunomagnetic nanoparticles [NPs], a microfluidic chip and an interdigitated microelectrode for impedance measurement. A polyclonal antibody against N1 subtype was immobilized on the surface of the microelectrode to specifically bind H5N1 virus. Therefore, chicken red blood cells were used as biolabels to attach H5N1 captured on the microelectrode to amplify impedimetric signals. A second antibody, also used against N1, offered greater specificity and reliability than the previous one. The results of this study showed that the biosensor was able to detect AIV in less than 2 hours (Lum & Li, 2010).

An optical interferometric waveguide immunoassay for direct and label-less detection of avian influenza virus is described by Xu et al., 2007. The assay is based on refraction index changes that occur upon binding of virus particles to antigen specific antibodies on the waveguide surface. Three virus subtypes were tested using both monoclonal and polyclonal capture antibodies. The detection limits were as low as 0.0005 HA units mL^{-1} (Xu et al., 2007). Liu et al. developed an electrochemical method for the detection of avian influenza virus using and aptamar of DeoxyriboNucleic Acid [DNA] as recognition element immobilized on a hybrid nanomaterial-modified electrode. The modified electrode was assembled with multi-walled carbon nanotubes [MWNT], polypyrrole nanowires [PPNWs] and AuNPs. The detection limit found was 4.3×10^{-13} M. These studies showed that the new hybrid nanomaterial MWNT/PPNWs/AgNPs and the DNA aptamer could be used to fabricate an electrochemical biosensor for gene sequence detection (Liu et al., 2011).

A spectroscopic assay based on Surface Enhanced Raman Scattering [SERS] using silver nanorod array substrates allowing the rapid detection of several human respiratory viruses with a high degree of sensitivity and specificity was developed by Shanmukh et al.. This novel SERS assay can detect spectral differences between viruses, viral strains and viruses with gene deletions in biological media. The method provides rapid detection and characterization of viruses generating reproducible spectra without viral manipulation (Shanmukh et al., 2006).

Persistent infection with hepatitis B virus [HBV] is a major health problem worldwide and may lead to chronic hepatitis, cirrhosis and primary liver cancer (Ye et al., 2003). The detection of HBV DNA in the serum of patient is becoming an important tool in the diagnosis of HBV infection. Two electrochemical detection methods based on DNA hydridization for detection of Hepatitis B viruses were reported by (Erdem et al., 2000 and Ye et al., 2003).

Varíola Virus, also call Smallpox virus [SpV], was irradicated in 1977 but with the increase of unprotected population this virus became an ideal warfare agent. Donaldson and co-workers developed a sensitive and rapid real time immunoassay to detect SpV (Donaldson et al., 2004). It consisted in a polystyrene optical waveguide coated with streptavidin where anti-Vaccinia antibody was attached as a biorecognition element. The detection was performed by a commercial single wavelength fluorometer designed for evanescent wave fluoroimmunoassays. The biosensor was able to detect a minimum of 2.5×10^5 pfu (pock-forming units)/ml of vaccinia virus in seeded throat culture swab specimens (Donaldson, 2004).

But other viruses have been detected rapidly by using biosensing technology. For example, immunosorbent electron microscopy was used to quantify recombinant baculovirus-generated bluetongue virus [BTV] core-like particles in either purified preparations or lysates of recombinant baculovirus-infected cells. The capture was the anti-BTV VP7 monoclonal antibody [MAb]. This technique is simple, rapid and accurate for the quantification of virus produced in large scale in vitro systems (Zheng et al., 1999).

Another example is the use of Atomic Force Microscopy [AFM], under dynamic conditions, for imaging single-stranded genomic RiboNucleic Acid [RNA] from four icosahedral viruses, in which the RNA was observed to unfold (Kuznetsov et al., 2005).

3. Bacteria biosensors

Bacteria are the major pathogen responsible for water-borne disease. Cholera, typhoid fever, bacillary dysentery, leptospirosis and gastroenteriris are some examples of waterborne diseases caused by *Vibrio cholera, Salmonella typhi, Shigella spp., Leptospira spp.,* and Enteropathogenic *Escherichia coli* [EPEC] (Ashbolt, 2004).

Table 1 resumes the latest screening methods for waterborne bacteria; capture and detection methods, limit of detection [LOD] and range of detection are compared (Baudart & Lebaron, 2010; Bharadwaj et al., 2011; Bruno et al., 2010; Chen et al., 2008; Duplan et al., 2011; Fu et al., 2010; Geng et al., 2011; Guven et al., 2011; Huang et al., 2008, 2011; Karsunke et al., 2009; Kwon et al., 2010; Li et al., 2011A; Luo et al., 2010; Miranda-Castro et al., 2009; Park et al. 2008; Sun et al., 2009; Wang et al., 2009; Wilbeboer et al., 2010; Wolter et al., 2008; Yoon et al., 2009; Yu et al., 2009; Xue et al., 2009).

Pathogens transmitted via water are mostly of faecal source (Ashbolt, 2004). Pathogenic *E. coli* strains are responsible for infection of the enteric, urinary, pulmonary and nervous systems. *E. coli* infection is usually transmitted through consumption of contaminated water or food. The routine detection methods for these microorganisms are based on Colony Forming Units [CFUs] count requiring selective culture, or biochemical and serological characterizations. Although bacterial detection by these methods is sensitive and selective, days are needed to get a result. Besides, these methods are costly and time consuming. Because of its great importance as faecal contaminant indicator in waters, the development of biosensors to detect and quantify *E. coli* has been extensively studied and there are a very large number of new methods and improvements to reference methods (Choi et al., 2007; Deobagkar et al., 2005; Gau et al., 2001; Yáñez et al., 2006; Liu & Li, 2002; Simpson & Lim 2005; Tang et al., 2006; Yoo et al., 2007; Yu et al., 2009).

Bacteria	Strain	Methods		LOD	Detection range	References
		Capture	Transduction			
E. coli	O157:H7	Immunological	Electrochemical	10^2 CFU mL^{-1}	10^2 - 10^5 CFU mL^{-1}	Yu et al., 2009
E. coli	O157:H7	Nucleic acid	Electrochemical	0.5 nmol L^{-1}	-	Wang et al., 2009
E. coli	ATCC 8739	Nucleic acid	Optical	30 CFU mL^{-1}	30 - 3×10^4 CFU mL^{-1}	Bruno et al., 2010
E. coli	unspecified	Nucleic acid	Electrochemical	50 cells mL^{-1}	1.0×10^2 - 2.0×10^3 cells mL^{-1}	Geng et al., 2011
E. coli	K12	Immunological	Optical	10^4 CFU mL^{-1}	10^2 - 10^6 CFU mL^{-1}	Duplan et al., 2011
E. coli	O157:H7	Immunological	Electrical	61 CFU mL^{-1}	0 - 10^4 CFU mL^{-1}	Luo et al., 2010
E. coli	O157:H7	Nucleic acid	Mechanical	1.2×10^2 CFU mL^{-1}	10^2 - 10^6 CFU mL^{-1}	Chen et al., 2008
E. coli	ATCC 15597	Immunological	Optical	10 CFU mL^{-1}	10 - 10^7 CFU mL-1	Yoon et al., 2009
E. coli	unspecified	Immunological	Optical	8 CFU mL^{-1}	10^1 - 10^4 CFU mL-1	Guven et al., 2011
E. coli	several	Nucleic acid	Optical	15 culturable E. coli (100 mL)$^{-1}$	3 5 and 14 5 viable E. coli (100 mL)$^{-1}$	Baudart & Lebaron, 2010
E. coli	DSM 30083	Enzimatic	Optical	7 CFU mL^{-1}	100 - 10^8 CFU mL^{-1}	Wildeboer et al., 2010
E. coli	ACTT 25922	Nucleic acid	Mechanical	8 Cells (800 mL)$^{-1}$	-	Sun et al., 2009
E. coli	K12	Immunological	Optical	100 CFU mL^{-1} (33% Ab coverage) 10 CFU mL^{-1} (50% and 100 % Ab coverage)	-	Kwon et al., 2010
E. coli	BD2399	Immunological	Mechanical	10^5 CFU mL^{-1}	5×10^4 - 5×10^9 CFU mL^{-1}	Fu et al., 2010
E. coli	DH5α	Nucleic acid	Electrochemical	5 CFU mL-1	1×10^3 - 5×10^5 CFU mL^{-1}	Li et al , 2011
E. coli	ATCC 35218	Immunological	Optical	< 1000 CFU mL-1	10^3 - 10^8 CFU mL^{-1}	Bharadwaj et al., 2011
E. coli	unspecified	TGA	Optical	10^2 CFU mL^{-1}	10^2 - 10^7 CFU mL^{-1}	Xue et al., 2009
E. coli	O157:H7	Immunological	Optical	1.8×10^3 CFU mL^{-1}	1.8×10^3 - 1.8×10^8 CFU mL^{-1}	Park et al., 2008
E. coli	O157:H7	Immunological	Optical	1×10^4 cells mL^{-1} (single) 3×10^3 cells mL^{-1} (multi)	1×10^4 - 1×10^6 cells mL^{-1} (single) 3×10^4 - 3×10^6 cells mL^{-1} (multi)	Wolter et al., 2008
E. coli	O157:H7	Immunological	Optical	1.8×10^4 cells mL^{-1}	8.5×10^4-3.7×10^6 cells ml^{-1}	Karsunke et al., 2009
E. coli	O157:H7	PU	Mechanical	2×10^2 cells mL^{71}	2×10^2- 3×10^6 cells ml^{72}	Huang et al., 2008
E. coli	K12 ER2925	Immunological	Electrical	10 cfu mL^{-1}	0 - 10^5 cfu mL^{-1}	Huang et al., 2010
Vibrio Chorela	O1	Immunological	Optical	1×10^3 CFU mL^{-1}	1×10^3 - 1×10^7 CFU mL^{-1}	Sungkanak et al., 2010
Vibrio Chorela	O1; O139	Immunological	Optical	10^8 CFU mL^{-1} (O1) (WOE) 10^7 CFU mL^{-1} (O139) (WOE) 10^2 CFU mL^{-1} (O1; O139) (WE)	10^1 -10^8 CFU mL^{-1}	Yu et al., 2011
Vibrio Chorela	several	Nucleic acid	Optical	10 CFU mL^{-1}	10 - 10^6 CFU mL^{-1}	Zhou et al., 2011
Leptospira interrogans	several	Nucleic acid	Optical	10 CFU mL^{-1}	10 - 10^6 CFU mL^{-1}	Zhou et al., 2011
Salmonella spp.	several	Nucleic acid	Optical	1 CFU mL^{-1}	10 - 10^6 CFU ml^{-1}	Zhou et al., 2011
Salmonella typhimurium	ATCC 14028	Immunological	Optical	3×10^6 cells mL-1 (single/multi)	3×10^6 - 3×10^8 cells mL^{-1} (single) 3×10^6 - 1×10^9 cells mL^{-1} (multi)	Wolter et al., 2008
Salmonella typhimurium	ATCC 14028	Immunological	Optical	2.0×10^7 cells mL^{-1}	5×10^6-1.1×10^9 cells mL^{-1}	Karsunke et al., 2009
Salmonella typhimurium	unspecified	Immunological	Mechanical	5×10^3 CFU mL^{-1}	5×10^3 - 5×10^8 CFU mL^{-1}	Guntupalli et al., 2007
Mycobacterium sp.	H37rv	Nucleic acid	Electrochemical	1.25 ng mL^{-1}	1.25 - 50 ng mL^{-1}	Thiruppathiraja et al., 2011

LOD – Limit Of Detection; WE – With Enrichment; WOE – WithOut Enrichment. ¿ CFU – Colony Form Unit; Ab – Antibody.

Table 1. Latest screening techniques for bacteria detection in waters.

The immunological methods are the most widely used as recognition methods (Bharadwaj et al., 2011; Duplan et al., 2001; Fu et al., 2010; Guven et al., 2011; Huang et al., 2011; Karsunke et al., 2009; Know et al., 2010; Luo et al., 2010; Park et al., 2008; Wolter et al., 2008; Yoon et al., 2009; Yu et al. 2009), but nucleic acid capture probe are starting to gain some importance in the field (Baudart & Lebaron, 2010; Bruno et al., 2010; Chen et al., 2008; Geng et al., 2011; Li et al., 2011; Sun et al., 2009; Wang et al., 2009). The use of aptamers instead of antibodies [Abs] as capture probes are increasing due to the advantages they present against Abs. Despite its high specificity and affinity, aptamers offer higher chemically stability and can be selected in vitro for a specific target, ranging from small molecules to large proteins and even cells. Moreover, once selected, aptamers can be synthesized with high reproducibility and purity from commercial sources. Furthermore, aptamers often undergo significant conformational changes upon target binding. This offers great flexibility in the design of novel biosensors with high detection sensitivity and selectivity (Song et al., 2008).

Bruno et al. described a Fluorescence Resonance Energy Transfer [FRET]-Aptamers assay applied to *E. coli* detection. In this assay, 25 reverse and 25 forward aptamer candidate sequences against *E. coli* were tested and compared in order to select the most sensitive to SPR and competitive-FRET analysis (Bruno et al., 2010). An iron oxide [Fe_2O_3] gold [Au] core/shell nanoparticle-based electrochemical DNA biosensor was developed for the amperometric detection of *E. coli* by Li et. al., 2011A The assay doesn't require any amplification step and the lowest detection value found was 5 CFU mL^{-1} of *E. coli* after 4.0 h of incubation (Li et al., 2011A). Another aptamer-based biosensor using magnetic particles was reported by Geng et al., 2011. This work included cobalt NPs and an enrichment process was necessary to find 10 *E. coli* cells mL^{-1} in real water samples by differential pulse voltammetry.

A flow piezoelectric biosensor based on synthesized thiolated probe specific to *E. coli* O157:H7 eaeA gene was immobilized onto the piezoelectric biosensor surface. Then, DNA hybridization was induced by exposing the immobilized probe to the *E. coli* O157:H7 eaeA gene fragment, resulting in a mass change and a consequent frequency shift of the piezoelectric biosensor. A second thiolated probe complementary to the target sequence was conjugated to the AuNPs to amplify the frequency change of the piezoelectric biosensor. The products amplified from concentrations of 1.2×10^2 CFU mL^{-1} of *E. coli* O157:H7 were detectable by the piezoelectric biosensor (Chen et al., 2008).

Sun et al. described a nano silver Indium Tin Oxide [ITO]-coated piezoelectric quartz crystal [PQC] electrode using DNA hybridization to detect *E. coli* cells. Neutravidin and a biotinylated probe were loaded at nano silver ITO-coated PQC and binding ratio was assessed. The binding ratio between the neutravidin and biotinylated DNA probe is increased from 1.00:1.76 of normal PQC to 1.00:3.01 using the nano-silver[Ag]-modified electrode, leading to an increase of more than 71% of the binding capacity of neutravidin to biotinylated DNA probes and an enhancement of 3.3 times for binding complementary DNA onto the nano-Ag-modified neutravidin/biotinylated DNA PQC biosensor. Under the optimized conditions, the detection limit was 0.4 ng/L for DNA Polymerase Chain Reaction [PCR] products or 8 *E. coli* cells in 800 mL in order to detect a single *E. coli* (Sun et al., 2009).

Magnetoelastic biosensors are based on the principle of change in the frequency of magnetoelastic materials (Nayak et al., 2009). Fu et al. developed a magnetostrictive microcantilever using physical absorption as detection system and an Ab against *E. coli* immobilized onto the surface of the microcantilever to form a biosensor. It was found a detection limit of 10^5 CFU mL^{-1} for a microcantilever with the size of 1.5mm×0.8mm×35μm (Fu et al., 2010).

Immuno-optical biosensors are largely the most detection systems used and many different approaches are developed. A poly(ethylene glycol) [PEG] hydrogel based microchip with patterned nanoporous aluminum oxide membrane [AOM] for bacteria fast patterning and detection with low frequency impedance spectrum was reported by Yu et al. The PEG hydrogel micropatterns on the saline-modified nanoporous alumina surface created controlled spatial distribution of hydrophobic and hydrophilic regions. These microwell arrays were composed of hydrophilic PEG sidewalls and hydrophobic silane-modified AOM bottom. Abs against *E. coli* were added and washed to form the patterns in the microwells. Then, the target bacteria was successfully patterned and captured inside the microwells. The microchip was able to detect bacteria concentrations of 10^2 CFU mL^{-1} (Yu et al., 2009).

Luo et al., 2010 developed a nitrocellulose nanofibers surface functionalized with anti-*E. coli* Abs on the top of silver electrodes Another Ab were coupled with conductive magnetic NPs and incubated with the test sample for target conjugation. After, the purified sample with the conductive label was dispensed on the application pad (a conductometric lateral flow biosensor). After capillary flow equilibrium, the direct-charge transfer between the electrodes was proportional to captured sandwich complex, which could be used to determine the pathogen concentration. The detection time of the biosensor was 8 min, and the detection limit 61 CFU mL^{-1} (Luo et al., 2010).

A microfluidic device with a portable spectrometer and UV to identify the signal intensity at 375 nm was developed by Know et al., 2010 and Yoon et al., 2009. The device fabricated by Kwon et al. was constructed in acrylic using an industrial-grade milling machine eliminating the need for photolithography and internal or external pumping. An automatic sampling system was built using drip emitters, such that the system could be connected to a pressurized water pipe for real-time detection of *E. coli* (Sungkanak et al., 2010). The microdevice developed by Yoon was fabricated by standard soft lithography with the poly(dimethyl siloxane) [PDMS]. A hole was made through the PDMS to make a view cell. Two cover glass slides were bonded to the top and bottom slides of a view cell using oxygen plasma asher. Two inlets and one outlet were then connected to Teflon tubing. A syringe pump was used to inject anti-*E. coli* conjugated latex particles and E. coli target solutions into the microchannel device (Yoon et al., 2009). Both devices presented 10 CFU mL^{-1} as detection limit.

A label-free technique based on evanescent wave absorbance changes at 280 nm from a U-bent optical fiber sensor was reported by Bharadwaj et al. Bending a decladded fiber into a U-shaped structure enhanced the penetration depth of evanescent waves and hence the sensitivity of the probe. A portable optical set-up with a UV Light-Emitting Diode [LED], a spectrometer and U-bent optical fiber probe of 200µm diameter, 0.75mm bend radius and effective probe length of 1 cm demonstrated an ability to detect less than 1000 CFU mL^{-1} (Bharadwaj et al., 2011).

A method combining immunomagnetic separation and SERS was developed by Guven et al., 2011. Au-coated magnetic spherical NPs were prepared by immobilizing biotin-labeled anti-*E. coli* Abs onto avidin-coated magnetic NPs and used in the separation and concentration of the *E. coli* cells. Raman labels have been constructed using rod shaped AuNPs coated with 5,5-dithiobis-(2-nitrobenzoic acid) [DTNB] and subsequently with a molecular recognizer. Then DTNB-labeled gold nanorods interacted with the gold-coated magnetic spherical nanoparticle-Ab-*E. coli* complex. The limit of detection value of this method was 8 CFU mL[-1].

The enzyme β-D-glucuronidase [GUD] is a specific marker for *E. coli* and 4-methylumbelliferone-β-D-glucuronide [MUG] a sensitive substrate for determining the presence of *E. coli* in a sample. Wildeboer et al., 2010, described a novel hand-held fluorimeter to directly analyse real samples for the presence of *E. coli*. The miniaturized fluorescence detector reduced the incubation time to 30 min and detect *E. coli* as low as 7 CFU mL[-1] in river water samples (Wildeboer et al., 2010).

The use of carbon allotropes like graphene is of a great potential in biosensing due to their extraordinary electrical, physical and optical properties. Huang et al. reported a graphene based biosensor to electrically detect *E. coli* bacteria with high sensitivity and specificity. The device has a graphene film immobilized with Abs against *E. coli* and a passivation layer. After exposure to *E. coli* bacteria, the graphene device conductance increased significantly. The lowest concentration detected was 10 CFU mL[-1]. The sensor also detected glucose induced metabolic activities of the bound *E. coli* bacteria in real time (Huang et al., 2011).

Some authors used synthetic materials as recognition materials (Huang et al., 2008; Xue et al., 2009). Xue et al. described a thioglycol acid [TGA] coupled with water-soluble quantum dots as a fluorescence marker for *E. coli* total count. TGA covalently bound to membrane protein of bacteria cells and after 30 min. fluorescence signal are clean seen at a fluorescent microscope (Xue et al., 2009). Huang et al. also reported a synthetic biosensor with polyurethane [PU] as capture probe of *E. coli* cells and a remote-query magnetoelastic system. The resonance frequency of a liquid immersed magnetoelastic sensor, measured through magnetic field telemetry, changed mainly in response to bacteria adhesion to the sensor and the liquid properties of the culture medium. During its growth and reproduction, *E. coli* consumed nutrients from a liquid culture medium that decreased the solution viscosity, and, in turn, changed the resonance frequency of the medium-immersed magnetoelastic sensor (Huang et al., 2008).

Cholera is described by WHO as an acute intestinal infection caused by ingestion of food or water contaminated with *Vibrio cholerae*. It has a short incubation period, from less than one day to five days, and produces an enterotoxin that causes a painless, watery diarrhea that can quickly lead to severe dehydration and death if treatment is not promptly given. Its severity may be confirmed by looking at Haiti recent outbreaks detected after the earthquake. Within 4 days of detecting an unusually high numbers of patients with acute watery diarrhea and dehydration, in some cases leading to death, the National Public Health Laboratory in Haiti isolated *Vibrio cholerae* serogroup O1, from patients in the affected areas of the earthquake. By March 2011, 4,672 people have died and 252,640 cases had been reported (Centers for Diseases Control and prevention [CDC], 2010). Now, cholera outbreaks are being reported since March 2011 along the Congo River, affecting the Democratic Republic of Congo and the Republic of Congo (WHO, 2011)…

Recently, an ultrasensitive microcantilever-based biosensor with dynamic force microscopy for the detection of *Vibrio cholera* O1 with a detection limit of 1×10^3 CFU mL[-1] and a mass sensitivity, $\Delta m/\Delta F$, of 146.5 pg/Hz was reported by Sungkanak et al. 2010. A dry-reagent gold nanoparticle-based lateral flow biosensor for the simultaneous detection of *Vibrio cholerae* serogroups O1 and O139, based on immunochromatographic principle, was also developed by (Yu et al., 2011).

Legionella pneumophila [L. pneumophila], which is associated with Pontiac fever and legionnaires disease, is commonly found in air conditioning and refrigerating towers, but can also be transported in drinking and freshwaters (Ashbolt, 2004). The detection of *L. pneumophila* in water samples using standard microbiological culture techniques is a delayed process because the bacterium is slow-growing and nutritionally fastidious, to the point that other species may compete with *Legionella* even when antibiotic are supplemented (Cooper et al., 2009).

Miranda-Castro et al. described an electrochemical sensing method for semiquantitative evaluation of *L. pneumophila* (Miranda-Castro et al., 2009). The DNA fragments were amplified by PCR and hybridized to a biotin-labeled reporter sequence and then to a thiolated stem-loop structure immobilized onto gold electrodes as a reporter molecule with 1-naphthyl phosphate as a substrate. 1-Naphthol enzymatically generated was determined by differential pulse voltammetry [DPV].

An immnuosensor using side-polished fiber based on SPR with halogens and LED light source (850 nm LED) for the detection of *L. pneumophila* was reported by (Lin et al., 2007). The SPR curve on the optical spectrum described on the optical spectrum analyzer [OSA] demonstrated a width wavelength of sensing range of SPR effects and sensitive responses. The detection limit was 10^1 CFU mL^{-1} and the detection range of 10^1 to 10^3 CFU mL^{-1} (Lin et al., 2007).

Optical Waveguide Lightmode Spectroscopy [OWLS] technique uses the evanescent field of a He-Ne laser which is coupled into a planar waveguide via an optical grating (Vörös et al., 2002). An OWLS real-time analytical system for the detection of *L. pneumophila* was described by Cooper et al.. An aqueous suspension of *L. pneumophila* was passed across the surface of waveguides functionalized with a specific anti-*Legionella* antibody. The binding between the bacterial cells and the Ab specific for that cell resulted in an increase in the refractive indices of the transverse electric and transverse magnetic photoelectric currents. The assay is presented as a rapid (25 min.) and sensitive (1.3×10^4 CFU mL^{-1}) detection method for *L. pneumophila* contamination in water samples (Cooper et al., 2009).

Regularly, water is contaminated with more than one pathogen, so the development of multi-analyte arrays are quiet relevant. Wolter et al. and Karsunke et al. reported a chemiluminescence microarray for the simultaneous detection of *E. coli*, *Salmonella typhimurium* and *L. pneumophila* in waters (Karsunke et al., 2009; Wolter et al., 2008).

The immuno-microarray described by Wolter et al. were produced on PEG-modified glass substrates by means of a contact arrayer. The chemiluminescence reaction was accomplished by a streptavidin-horseradish peroxidase catalyzed reaction of luminol and hydrogen peroxide, and was recorded by a sensitive charge coupled device [CCD] camera. The detection limits, in multianalyte experiments, achieved in 13 minutes, were 3×10^6 cells mL^{-1}, 1×10^5 cells mL^{-1}, and 3×10^3 cells mL^{-1} for *S. typhimurium*, *L. pneumophila*, and *E. coli* O157:H7, respectively (Wolter et al., 2008).

Zhou et al. reported the development of a DNA microarray for detection and identification of *L. pneumophila* and ten other pathogens like *Salmonella spp.*, *Leptospira interrogans* and *Vibrio Cholera*, in drinking water. The combined two-step enrichment procedure allows detection sensitivity of 0.1 ng DNA or 10^4 CFU mL^{-1} achieved for pure cultures of each target organism (Zhou et al., 2011).

Salmonella is related with water and food-borne severe diarrhea. It contaminates more often dairy products from poultry but because these pathogens can be present in animal faeces water resources can as well be contaminated (Karsunke et al., 2009; Nayak et al., 2009; Wolter et al., 2008; Zhou et al., 2011).

Guntupalli et al. developed a magnetoelastic biosensor for *S. typhimurium* in a mixed microbial population. A varying magnetic field was applied to magnetoelastic particles coated with an Au/Chromium [Cr] thin film and a specific Ab in contact with the analyte and the resulting signal was measured. A detection limit of 5×10^3 CFU mL^{-1} and the sensitivity of 139 Hz decade^{-1} were observed for a $2\times0.4\times0.015$ mm sensor (Guntupalli et al., 2007).

About 90 species of Mycobacteria have been found, 20 of which are known to cause disease in humans. Among these, *M. avium, M. intracellulare, M. chelonae, M. kansasii, M. marinum, M. fortuituma and M. ulcerans* are considered potential human pathogens. These are called Non-tuberculosis mycobacteria [NTM] and have been identified in numerous environmental sources, including water. NTM species have been isolated from water sources as well as waste water, surface water, recreational water, ground water and tap water. In this field, there are some recent advances in the development of fast-screening method to detect Mycobacteria. Hunter et al. published a research report about treatment, distribution, standard methods isolation, comparison of culture media and analysis in water (Hunter et al., 2001). Most recent developments in these area report PCR-based detection methods (Jing et al., 2007; Tobler et al., 2006), and demand new emerging biosensing technology.

Recent efforts for detecting whole bacterial communities have been made in order to detect possible bacterial pathogens which are not included in the standard monitoring processes (Hong et al., 2010; Kormas et al., 2010; Poitelon et al., 2010; Revetta et al., 2010). The analysis of bacterial 16S rRNA gene diversity in drinking water distribution systems [WDS] can indicate the presence of Proteobacteria even after the chlorine disinfection treatment (Revetta, 2010).

Poitelon and coworkers presented the variations of 16S rDNA phylotypes prior and after chlorination treatments in 2 WDS in France. The 16S rDNA sequences were grouped into operational taxonomic units [OTUs] standards and acquired for each sample. Significant differences were found in terms of structure and composition of the bacterial community before and after the chlorination disinfection. The predominant bacteria found were alpha- and betaproteobacteria (Poitelon et al., 2010). Kormas, 2010, also presented a similar study in the WDS of Trikala city, Greece. This work showed a possible microbiological risk, not from microorganisms that are routinely monitoring, but from Mycobacteria-like Bacteria.

Overall, the wiseness of the ecology of the bacterial communities in the WDS is important to be acquainted with the chlorination resistance of pathogen in these microbiological communities (Kormas et al., 2010).

4. Cyanobacteria and algae biosensors

The toxic metabolites produced by *Cyanobacteria* in surface waters are another potential threat to drinking water. This has been recognized recently after detecting cyanobacterial toxins [cyanotoxins] in a great number of water samples, from nearly every region on earth. Outbreaks of human poisoning attributed to toxic *Cyanobacteria* have been reported, following exposure of individuals to contaminated water, while drinking, swimming and canoeing (WHO, 1999).

Cyanotoxins [CNs] belong to rather diverse groups of chemical substances each of which shows specific toxic mechanisms in vertebrates. Some CNs are strong neurotoxins (anatoxin-a, saxitoxins), others are primarily toxic to the liver (microcystins, nodularin and cylindrospermopsin), and yet others (such as the lipopolysaccharides) appear to cause gastroenteritis (WHO, 1999). Microcystins [MCs] are geographically the most widely CNs distributed in freshwaters. MCs are peptide toxins produced by *Cyanobacteria* populations in high proportions, a typical symptom of eutrophication. Microcystin-LR [MC-LR] is the most typical element of this group and seems to display severe hepatotoxic and carcinogenic activity. MC-LR concentration in waters for human consumption is regulated by environmental Protection Agency [EPA], EU and WHO, who recommended a limit of 1 µg L^{-1} MC-LR in drinking waters (EC, 2000, 2006; WHO, 1999). Current standard methods to monitor MC-LR require sophisticated and expensive procedures and specific laboratorial conditions that take long time to reach the intend result. Within that time the contamination can move/spread out, while users of recreational waters and possible consumers are at risk of contracting serious infections.

A huge number of novel methods and techniques have been developed recently for this purpose. They are presented in table 2 (Almeida et al., 2006; Allum et al., 2008; Campàs et al., 2007; Campàs & Marty, 2007; Dawan et al., 2011; Ding & Mutharasan, 2010; Gregora & Marsálek, 2005; Hu et al., 2008; Lindner et al., 2009; Long et al., 2009, 2010; Loypraseta et al., 2008; Pyo et al., 2005, 2006; Pyo & Jin, 2007; Queirós et al., 2011; Sheng et al., 2007; Wang et al., 2009; Xia et al., 2010, 2011; Zhang et al., 2007). Optical immuno-based techniques are by large the most widely techniques that have been developed for the detection of MCs (Campàs et al., 2007; Hu et al., 2008; Long et al., 2010; Loypraseta et al., 2008; Pyo et al., 2005, 2006; Pyo & Jin, 2007; Sheng et al., 2007; Xia et al., 2010).

The chemiluminescence sensing technique used in microbial biosensors relies on the generation of electromagnetic radiation as light by the release of energy from a chemical reaction using synthetic compounds (highly oxidized species) which respond to the target analyte in a dose-dependent manner (Hu et al., 2008; Lindner et al., 2009; Su et al., 2011). Lindner et al. reported a rapid immunoassay for sensitive detection of MC-LR using a portable chemiluminescence multichannel immunosensor. The sensor device is based on a capillary ELISA technique in combination with a miniaturized fluidics system and uses chemiluminescence as the detection principle. Minimum concentrations of 0.2 µg L^{-1} MC-LR could be measured in spiked buffer as well as in spiked water samples (Wang et al., 2009). Hu et al. also presented a chemiluminescence immunosensor based on gold nanoparticles. The immunoassay included three main steps: indirect competitive immunoreaction, oxidative dissolution of AuNPs, and indirect determination for MCs with Au^{3+}-catalysed luminol chemiluminesent system. The method has an extensive working range of 0.05–10 µg L^{-1}, and a limit of detection of 0.024 µg L^{-1} (Hu et al., 2008).

Evanescent wave fiber optic immunosensors have been developed to determine numerous target compounds based on the principle of immunoreaction and total internal reflect fluorescent. They also show potential advantages such as miniaturization, sensitivity, cost-effectiveness, and capability of real-time rapid measurements (Long et al., 2009). A portable trace organic pollutant analyzer based on the principle of immunoassay and total internal reflection fluorescence was developed by Long et al.. The reusable fiber optic probe surface was produced by covalently immobilizing a MC-LR-ovalbumin conjugate onto a self-assembled thiol-silane monolayer of fiber optic probe through a heterobifunctional reagent. The recovery of MC-LR added to water samples at different concentrations ranged from 80 to 110% with relative standard deviation values less than 5% (Long et al., 2009).

Pyo et al. developed a few number of optical immuno-based assays to detect MC-LR: a gold colloidal immunochromatographic strip, a fluorescence immunochromatographic strip and cartridge and a PDMS microchip using a liquid-cord waveguide as optical detectors and a MAb against MC-LR as recognition element. The detection limits found were 0.05 µg L^{-1}, 0.15 µg L^{-1} and 0.05 µg L^{-1}, respectively (Pyo et al., 2005, 2006; Pyo & Jin, 2007).

Optical sensing techniques, and specially integrated in immunoarrays, are particularly powerful tools in high throughput screening to monitoring multiple analytes simultaneously (Long et al., 2010; Su et al., 2011). Long et al. developed an optical fiber-based immunoarray biosensor for the detection of multiple small analytes (MC-LR and trinitroluene [TNT]). These compounds can be detected simultaneously and specifically within an analysis time of about 10 min for each assay cycle. The limit of detection for MC-LR was 0.04 µg L^{-1}. This compact and portable immunoarray shows good regeneration performance and binding properties, robustness of the sensor surface and accuracy in the measurement of small analytes which can be considered an excellent multiple assay platform for clinical and environmental samples (Long et al., 2010).

A suitable disposable type biosensor for on-site monitoring of MC-LR in environmental waters was described by Zhang et al.. They reported a competitive binding non-separation electrochemical enzyme immunoassay, using a double-sided microporous gold electrode in cartridge-type cells. Mean recovery of MC-LR added to tap water was 93.5%, with a coefficient of variation of 6.6% (Zhang et al., 2007).

Electrochemical detection methods are also commonly used (Campàs et al., 2007; Campàs & Marty, 2007; Dawan et al., 2011; Loypraseta et al., 2008; Pyo & Jin, 2007, Wang et al., 2009; Zhang et al., 2007).

Campàs et al. developed a highly sensitive using both MAb and Polyclonal Ab [PAb] against MC-LR. This amperometric immunosensor was compared with colorimetric immunoassay, PP inhibition assay and High Performance Liquid Chromatography [HPLC]. The amperometric immunosensor simplifies the analysis, and offer faster and cheaper procedures than others assays.

Wang et al. described a simple, rapid, sensitive, and versatile ELISA-SWNT-paper based sensor for detection of MC-LR in waters. This paper described the use of paper saturated with SWNT embebed in a poly(sodium 4-styrenesulfonate) solution where the Abs are after immobilized. The chronoamperometric detection system was tested and performed a LOD of 0.6 µg L^{-1} with a linear range up to 10 µg L^{-1} (Wang et al., 2009).

Methods		Microcystins	LOD $(\mu g\ L)^1$	Detection range $(\ \mu g^{-1}L^{-1}$	References
Capture	Transduction				
Immunological	Electrochemical	-LR	0.6	up to 10	Wang. et al., 2009
Immunological	Optical	-LR	0.03	0.1-10	Long et al., 2009
Immunological	Optical	-LR -RR	down to 0.1	0.001-30	Sheng et al., 2007
Immunological	Piezoelectric	-LR	0.1 (river) 1.00E^3 (tap)	1-100	Ding & Mutharasan, 2010
Immunological	Optical	Any	0.024	0.05–10	Hu et al., 2008
Molecular Imprinting	Optical	-LR	0.3	0.3-1.4	Queirós et al., 2011
Enzimatic	Optical	-LR	1.00E-02	0-100	Almeida et al., 2006
Immunological	Optical	-LR	1.50E-01	0.150-1.6	Pyo & Jin, 2007
Immunological	Electrochemical	-LR	0.10 (MAb) 1.73 (PAb)	10^4-10^2 (MAb) 10^{71}-10^2 (PAb)	Campàs & Marty, 2007
Immunological	Mechanical	-LR	1	1-1000	Xia et al., 2011
Enzimatic	Optical	-LR	1	0-1000	Allum et al., 2008
Immunological	Optical	-LR	0.2	0-10	Lindner et al., 2009
Enzimatic	Electrochemical Optical	Any	37 (Electrochemical) 2 (Optical)	37-188 (Electrochemical) - (Optical)	Campàs et al., 2007
Immunological	Electrochemical	-LR	0.1	0.01-3.16	Zhang et al., 2007
Immunological	Electrochemical	-LR	7.00E-06	10E^6-1	Loypraserta et al., 2008
Immunological	Optical	-LR	1	1-100	Xia et al., 2010
Immunological	Optical	-LR	0.05	0.05-1.2	Pyo et al., 2005
Immunological	Optical	-LR	0.05	0-1.6	Pyo et al., 2006
Immunological	Optical	-LR	0.04	0-1000	Long et al., 2010
Immunological	Electrochemical	-LR	1.00E-06	1.00E^6-1.00E^4	Dawan et al., 2011

LOD – Limit Of Detection; WE – With Enrichment; WOE – WithOut Enrichment.

Table 2. Emerging techniques for the detection of microcystins in waters.

The use of electrochemical capacitance systems as detection methods in biosensors allows obtaining exceptionally low limits of detection. Loypraserta et al. and Dawan et al. described a label-free immunosensor based on a modified gold electrode incorporated with AgNPs to enhance the capacitive response to MC-LR has been developed. Anti-MC-LR was immobilized on AgNPs bound to a self-assembled thiourea monolayer, and compared with a bare electrode without the modified AgNPs (Dawan et al., 2011; Loypraseta et al., 2008). MC-LR could be determined with a detection limit of 7.0 and 1.0 µg L^{-1} for (Loypraseta et al., 2008) and (Dawan et al., 2011), respectively. Compared with the modified electrode without AgNPs, this assay presented higher sensitivity and lower limit of detection.

The Adda (3-amino-9-methoxy-2,6,8-trimethyl-10-phenyldeca-4,6-dienoic acid) group of MCs is responsible by the infiltration of toxins in the liver cells where MCs irreversibly bind to serine/threonine protein phosphatases type 2A [PP2A] and type 1 [PP1], inhibiting their enzymatic activity (Almeida et al., 2006; Campàs et al., 2007). These enzymes are involved on the dephosphorylation of proteins. Consequently, their inhibition results in hyperphosphorylation and reorganisation of the microfilaments, promoting tumours and liver cancer (Almeida et al., 2006; Campàs et al., 2007). Almeida et al., Allum et al. and Campàs et al. developed enzymatic-based capture probes based on PP1 and PP2A inhibition. Almeida et al. and Allum et al. presented a simple, rapid and reproducible PP1 inhibition colorimetric test, with the first one presenting a detection limit of $1.00E^{-2}$ µg L^{-1}. Allum et al. showed an optical fluorometric biosensor based on protein phosphate for the detection of MC-LR or diarrhetic shellfish toxins. The immobilised format described was used to evaluate the potential to translate protein phosphate into a prototype biosensor suitable as a regulatory assay allowing faster throughput than a solution assay and in particular, in assessments of sensitivity and reusability (Allum et al., 2008).

A chronoamperometric biosensor was developed by Campàs et al. 2007 also based on the inhibition of the PP2A. The enzyme was immobilized by the use of poly(vinyl alcohol) azide-unit pendant water-soluble photopolymer. The standard inhibition curve has provided a 50% inhibition coefficient [IC50] of 83 µg L^{-1}, which corresponds to a limit of detection of 37 µg L^{-1} (35% inhibition). Real samples were analysed using the developed amperometric biosensor and compared to those obtained by a conventional colorimetric protein phosphatase inhibition [PPI] assay and HPLC. The results demonstrated that the developed amperometric biosensor may be used as screening method for MCs detection (Campàs et al., 2007).

A dendritic surfactant for MC-LR detection by double amplification in a QCM biosensor was presented by Xia et al.. For primary amplification, an innovative interface on the QCM was obtained as a matrix by the vesicle layer formed by a synthetic dendritic surfactant. The vesicle matrix was functionalized by a MAb against MC-LR to detect the analyte. The results showed that a detection limit of 100 µg L^{-1} was achieved by the first amplification. A secondary amplification was implemented with anti-MC-LR AuNPs conjugates as probes, which lowered the detection limit for MC-LR to 1 µg L^{-1} (Xia et al., 2011).

A piezoelectric-excited millimeter-sized cantilever sensor was developed by Ding et al. for the sensitive detection of MC-LR in a flow format using MAb and PAb that bind specifically to MC-LR. Monoclonal antibody against MC-LR was immobilized on the sensor surface via amine coupling. As the toxin in the sample water bound to the antibody, resonant frequency decreased proportional to toxin concentration. Positive verification of MC-LR detection was confirmed by a sandwich binding on the sensor with a second antibody binding to MC-LR on the sensor which caused a further resonant frequency decrease (Ding &_Mutharasan, 2010).

Very recently, Queirós et al. described a completely new approach a Fabry–Pérot sensing probe based on an optical fibre tip coated with a MC-LR selective thin film. The membranes were developed by molecular imprinting of MC-LR in a sol–gel matrix that was applied over the tip of the fibre. The sensor showed low thermal effect, thus avoiding the need of temperature control in field applications. It presented a detection limit of 0.3 µg L^{-1} and shows excellent selectivity for MC-LR against other species co-existing with the analyte in environmental waters (Queirós et al., 2011).

Saxitoxins were originally isolated from shellfish where they are concentrated from marine dinoflagellates and are one of the causative agents of paralytic shellfish poisoning [PSP] and have caused deaths in humans (Haughey et al., 2011; WHO, 1999; Yakes et al., 2011).

Saxitoxins have been found in the cyanobacteria *Aphanizomenon flos-aquae*, *Anabaena circinalis*, *Lyngbya wollei* and *Cylindrospermopsis raciborskii*. Saxitoxins have been found in some countries in diverse cyanobacteria genera, such as *Aphanizomenon flos-aquae* strains NH-1 and NH-5 in North America, C1 and C2 toxins from *Anabaena circinalis* strains in Australia and *Cylindrospermopsis raciborskii* in Brazil (WHO, 1999).

Institute of Agri-Food and Land Use, the US Food and Drug Administration, and the Joint Institute for Food Safety and Applied Nutrition recently reported several SPR platforms biosensor based on inhibition assays to detect PSP toxins (Haughey et al., 2011; Yakes et al., 2011). A saxitoxin PAb (R895) and a MAb (GT13A) were tested and compared. MAb GT13A shows a higher sensitivity 77.8%-100% and was then immobilized into the SPR biosensor surface. The final system provides rapid substrate formation, about 18 h for saxitoxin conjugation with low reagent consumption, contains a reference channel for each assay, and is capable of triplicate measurements in a single run with detection limits well below the regulatory action level (Haughey et al., 2011; Yakes et al., 2011).

Other biosensors were developed for detection of other toxic algae, beyond cyanobacteria, present in environmental waters. Orozco et al. described an electrochemical DNA-based sensor device for detecting toxic algae (*Prymnesium parvum, and Gymnodinium catenatum*). A sandwich hybridization assay was developed with a thiol (biotin) labelled capture probe immobilized onto gold or carbon electrodes, the synthetic DNA was applied to the sensor and allowed to hybridize to the capture probe. A signal probe with HRP label was then applied, followed by an antibody to the HRP and a substrate. The electrical signal obtained from the redox reaction was proportional to the amount of DNA applied to the biosensor (Orozco et al., 2011; Orozco & Medlin, 2011).

Diercks-Horn et al. reported very recently the ALGADEC device which is a semi-automated ribosomal RNA [rRNA] biosensor for the detection of *Alexandrium minutum* toxic algae. The biosensor consisted of a multiprobe chip with an array of 16 gold electrodes for the detection of up to 14 target species. The multiprobe chip was placed inside an automated hybridization chamber, which was in turn placed inside a portable waterproof case with reservoirs for different reagents. The device processed automatically the main steps of the analysis and completed the electrochemical detection of toxic algae in less than 2 h (Diercks-Horn et al., 2011). Furthermore, Diercks et al. reported a colorimetric sandwich hybridization in a microtiter plate assay also for the detection of *Alexandrium minutum*. The system is an adaptation of a commercially available PCR ELISA Dig Detection Kit to be possible the rapid assessment of specificity of the two probes. The mean concentration of RNA per cell of was determined to be 0.028 ng±0.003 (Diercks et al., 2008).

Metfies et al. described an electrochemical DNA-biosensor for the detection of the toxic dinoflagellate *Alexandrium ostenfeldii* (Metfies et al., 2005). This device is similar to the one reported by Diercks-Horn et al. very recently.

5. Fungi biosensors

The knowledge-based concerning the occurrence of fungi in water is low. Fungal contamination of water has been reported for decades, although investigations have been inadequate compared with those of bacteria. Fungi are responsible by infections, allergic reactions and production of toxins [mycotoxins]. Mycotoxins contaminate food and drinks with harsh effects on human and animal health, including cancer, immunological effects and death (Russell et al., 2005). Water, either drinking or nondrinking, may be an effective medium for toxin and biological weapon dispersal. In nondrinking water, the toxin could be spread, from a shower and then inhaled. Workplaces such as farms or car washes where high volumes of water are employed could be susceptible to toxins or fungi. Furthermore, drinking water for animals may be at a considerably higher level of risk than that for humans, thereby increasing the threat to food sources (Russell et al., 2010).

Russell and Paterson proved the production of the mycotoxin zearalenone [ZEN] in drinking waters by *Fusarium graminearum* (Russell & Paterson, 2007). ZEN was purified with an immunoaffinity column and quantified by HPLC with fluorescence detection. The extracellular bear of ZEN was 15.0 ng L^{-1} (Russell et al., 2005; Russell & Paterson, 2007).

Aflatoxins, from Aspergillus flavus, were detected from a cold water storage tank and first reported as the first reported natural occurrence of any mycotoxin in waters (Paterson et al., 1997). Very recently Kattke et al. reported a FRET-based quantum dot immunoassay for rapid and sensitive detection of *Aspergillus amstelodami* (Kattke et al., 2011). The biosensor complex is formed when a quencher-labeled analyte is bound by the antigen-binding site of the quantum dot-conjugated antibody; when excited, the quantum dot will transfer its energy through FRET to the quencher molecules due to their close proximity. With the addition of the target analytes the quencher-labeled analytes dislocation causes disruption of FRET, which translates to increased quantum dot donor emission signal. The optimized displacement immunoassay detected *A. amstelodami* concentrations as low as 10^3 spores mL^{-1} in less than five minutes (Kattke et al., 2011).

Few biosensors for the rapid detection of different mycotoxins in food and beverages samples with special focus to electrochemical detection methods (Alonso-Lomillo et al., 2010; Dinçkaya et al., 2011; S. Li et al., 2011) and enzymes as capture probes (Alonso-Lomillo et al., 2010; S. Li et al., 2011) have been developed recently.

Li et al. presented an amperometric aflatoxin B1 biosensor developed by aflatoxin-oxidase [AFO], embedded in sol-gel, linked to MWCNTs-modified platinum [Pt] electrode (Li et al., 2011B). A DNA-based gold nanoparticles biosensor was developed for detection of aflatoxin M1. A self-assembled monolayer of cysteamine and gold nanoparticles on the SAM were prepared on gold electrodes, layer-by-layer. The assembly processes of cysteamine, gold nanoparticles, and ss-HSDNA were monitored with the help of electrochemical impedance spectroscopy [EIS] and cyclic voltammetry [CV] techniques. The biosensor provided a linear response to aflatoxin M1 over the concentration range of 1–14 ng mL^{-1} with a standard deviation of ±0.36 ng mL^{-1} (Dinçkaya et al., 2011).

Biosensors for the rapid detection of Ochratoxin [Ochra] were also described. Ochra is a group of mycotoxins produced as secondary metabolites by fungi which presents a serious hazard to human and animal health (Metfies et al., 2005; Russell et al., 2010; Yuan et al., 2009).

A sensitive enzyme-biosensor based on screen-printed electrodes was presented by Alonso-Lomillo et al., 2010. A competitive immunoassay with optical detection (SPR using AuNPs) was described by Yuan et al., in whom a mixed self-assembled monolayer was arranged with the immobilization of Ochra through its ovalbumin conjugated with a polyethylene glycol linker (Yuan et al., 2009).

Sapsford et al. reported an array biosensor for the detection of multiple mycotoxins such as ochratoxin A, fumonisin B, aflatoxin B1 and deoxynivalenol in food or beverages. The arrangement of the test was a competitive-based immunoassay, using monoclonal antibodies and the simultaneous detection was performed in less than 15 minutes (Sapsford et al., 2006).

6. Protozoa biosensors

Further on, it is estimated that 2.5 million cases of world annual illness outcome from food and water parasites. Four parasites, *Cryptosporidium parvum, Cyclosporacayetanensis, Giardia lamblia* and *Toxoplasma gondii,* account for the majority of cases, with 71% of waterborne diseases caused by *G. lamblia* and *Cryptosporidium* (Ashbolt, 2004; Rasooly & Herold, 2006; Smith & Nichols, 2010).

Frequent methods for protozoa detection are PCR-based assays (Loge et al., 2002; Toze, 1999) and more recently immunomagnetic separation and fluorescent assays (Ferrari et al., 2006; Mons et al., 2009). These techniques are used to evaluate the microbiological quality of water, and to assess the efficiency of pathogen removal in drinking and wastewater treatment plants (Girones et al., 2010).

Li et al., 2011B, reported a most likely biosensor approached for the detection of *G. lamblia* cysts based on the catalytic growth of AuNPs (X.X. Li et al., 2009). The assay can be described in 7 steps. First the sample is transferred and secondly concentrated; the third step is the incubation of the AuNPs immobilized with a MAb against *G. lamblia* as capture probe, the free gold probes are separated and the sample are resuspended and left growing; the seventh step is the detection by UV-VIS. Detection limit of *G. lamblia* cysts was determined as low as 1.088×10^3 cells ml^{-1} (X.X. Li et al., 2009).

7. Conclusion

Although WHO considers H5N1 water contamination an emerging issue, with high potential risks of infection, it seems that most issued waterborne viruses include EV. The toxic metabolites produced by *Cyanobacteria* in surface waters are another potential threat to drinking water, being MC-LR widely studied and target of biosensing development. Despite the low occurrence of fungi in water, mycotoxins have been also reported and target of biosensing technology. *Bacteria* are definitely the major pathogenic specie responsible for water-borne disease. They are mostly from faecal source. *E. coli* has been assayed thoroughly by immunological and nucleic acid methods, coupled to electrochemical or optical detection methods. *Protozoa* can be 0.01 mm to 1.0 mm in *length* and are quite easy to identify, turning out biosensors less desired. They are much less reported and detected mostly by PCR-based assays, which are used to evaluate the microbiological quality of water.

The biosensor schemes for detection of pathogens in waters presented in this chapter are only a few of many emerging devices and assays that are being developed for several applications. Biosensors offer high sensitivity and specificity, allowing the detection of extremely low infection doses and multiple target analyte at once. The integration of several sensors in networks providing multi-analyte detection and their portability of these sensor networks are at the forefront. This allows the fast screening and the prevention of outbreaks and its consequences.

The future directions of the use of biosensors in environmental contaminant analysis may lead to the combination of these sensor networks with wireless signal transmitters for remote sensing at real-time monitoring. Low manufacturing and production costs are also attractive features that may lead the market expansion in a very short time. They are indeed promising tools for field detection and fast screening of pathogens in waters...

8. Nomenclature

Ab	Antibody
Abs	Antibodies
Adda	3-amino-9-methoxy-2,6,8-trimethyl-10-phenyldeca-4,6-dienoic acid
AFM	Atomic Force Microscopy
AFO	Aflatoxin-oxidase
Ag	Silver
AgNPs	Silver NanoParticles
AIV	Avian Influenza Virus
AOM	Aluminium Oxide Membrane
ATR	Attenuated Total Reflectance
Au	Gold
BTV	BlueTongue Virus
CCD	Charge Coupled Device
CFU	Colony Form Unit
CNs	Cyanotoxins
Cr	Chromium
CV	Cyclic Voltammetry
DNA	DeoxyriboNucleic Acid
DPV	Differential Pulse Voltammetry
DTNB	5,5-dithiobis-(2-nitrobenzoic acid)
E. Coli	Escherichia coli
EIS	Electrochemical Impedance Spectroscopy
ELISA	Enzyme-Linked ImmunoSorbant Assay
EPA	Environmental Protection Agency
EPEC	EnteroPathogenic Escherichia coli
EU	European Union
EV	Enteric Viruses
FETs	Field Effect Transistors
FRET	Fluorescence Resonance Energy Transfer
GUD	b-D-glucuronidase
HBV	Hepatitis B Virus

HPAI	Highly Pathogenic Avian Influenza
HPLC	High Performance Liquid Chromatography
HRP	HorseRadish Peroxidase
IC50	half maximal Inhibitory Concentration
IR	Infra-Red
ITO	Indium tin oxide
LED	Light-Emitting Diode
LOD	Limit Of Detection
MAb	Monoclonal Antibody
MC-LR	Microcystin-LR
MCs	Microcystins
MOS	Metal Oxide Semiconductor
MUB	4-methylumbelliferone-b-D-glucuronide
MWNT	Multi-Walled carbon NanoTubes
NMR	Nuclear Magnetic Ressonance
NPs	Nanoparticles
NTB	Non-Tuberculosis Mycobacteria
NWs	NanoWires
Ochra	Ochratoxin
OSA	Optical Spectrum Analyzer
OWLS	Optical Waveguide Lightmode Spectroscopy
PAb	Polyclonal Antibody
PCR	Polymerase Chain Reaction
PEG	Poly(ethylene glycol)
Pfu	Pock-forming units
PMDS	Polydimethylsiloxane
PP	PolyPyrrole
PP1	Protein Phosphatase 1
PP2A	Protein Phosphatase 2A
PPI	Protein Phosphatase Inhibition
PPNWs	PolyPyrrole NanoWires
PQC	Piezoelectric Quartz Crystal
PSP	Paralytic Shellfish Poisoning
Pt	Platinum
PU	Polyurethane
QCM	Quartz Crystal Microbalance
RNA	RiboNucleic Acid
rRNA	ribosomal RNA
RT-PCR	Reverse Transcriptase – Polymerase Chain Reaction
SAW	Surface Acoustic Wave
SERS	Surface Enhanced Raman Scattering
SPR	Surface Plasmon Resonance
SpV	Smallpox Virus
SWNT	Single-Wall NanoTube
TGA	Thioglycol Acid
TNT	Trinitroluene
UV	Ultra-Violet

Vis	Visible
WHO	World Health Organization
WISE	Water Information System for Europe
ZEN	Zearalenone

9. References

Allum, L.L., Mountfort, D.O., Gooneratne, R., Pasco, N., Goussaind G., & Halle, E.A.H. (2008). Assessment of protein phosphatase in a re-usable rapid assay format in detecting microcystins and okadaic acid as a precursor to biosensor development. *Toxicon 52 (7)*, pp. 745-753.

Almeida, V., Cogo, K., Tsai, S.M., & Moon, D.H. (2006). Colorimetric test for the monitoring of microcystins in Cyanobacterial culture and environmental samples from southeast – Brazil. *Brazilian Journal of Microbiology 37*, pp. 192-198.

Alonso-Lomillo, M.A., Domínguez-Renedo, O., Ferreira-Gonçalves L., & Arcos-Martíneza, M.J. (2010). Sensitive enzyme-biosensor based on screen-printed electrodes for Ochratoxin A. *Biosensors and Bioelectronics 25 (6)*, pp. 1333-1337.

An, J., & Carmichael, W.W. (1994). Use of colorimetric protein phosphatase inhition assay and enzime linked immunosorbent assay for the study of microcystins and nodularin. *Toxicon 32*, pp. 1495-1507.

Ashbolt, N.J. (2004). Microbial contamination of drinking water and disease outcomes in developing regions. *Toxicology 198*, pp. 229-238.

Bao, P.-D., Huang, T.-Q., Liu, X.-M., & Wu, T.-Q. (2001). Surface-enhanced Raman spectroscopy of insect nuclear polyhedrosis virus. *J. Raman Spectrosc. 32*, pp. 227-230.

Barenfanger, J., Drake, N., Leon, N., Mueller, T., & Troutt, T. J. (2000). Clinical and Financial Benefits of Rapid Detection of Respiratory Viruses: an Outcomes Study. *Clin. Microbiol. 38*, pp. 2824-2828.

Baudart, J., & Lebaron, P. (2010). Rapid detection of *Escherichia coli* in waters using fluorescent in situ hybridization, direct viable counting and solid phase cytometry. *Journal of Applied Microbiology 109(4)*, pp. 1253–1264.

Bharadwaj, R., Sai, V.V.R., Thakare, K., Dhawangale, A., Kundu, T., Titus, S., Verma, P.K., & Mukherji, S. (2011). Evanescent wave absorbance based fiber optic biosensor for label-free detection of *E. coli* at 280nm wavelength. *Biosensors and Bioelectronics 26*, pp. 3367–3370.

Brassard, J., Seyer, K., Houde, A., Simard, C., & Trottier, Y.-L. (2005). Concentration and detection of hepatitis A virus and rotavirus in spring water samples by reverse transcription-PCR. *Journal of Virological methods 123*, pp. 163-169.

Bruno, J.G., Carrillo, M.P., Phillips, T., & Andrews, C.J. (2010). A Novel Screening Method for Competitive FRET-Aptamers Applied to E. coli Assay Development. *J Fluoresc. 20*, pp. 1211–1223.

Campàs, M., & and Marty, J.-L. (2007). Highly sensitive amperometric immunosensors for microcystin detection in algae. *Biosensors and Bioelectronics 22*, pp. 1034–1040.

Campàs, M., Szydłowska, D., Trojanowicz, M., & Marty, J.-L. (2007). Enzyme inhibition-based biosensor for the electrochemical detection of microcystins in natural blooms of cyanobacteria. *Talanta 72 (1)*, pp. 179-186.

Centers for Diseases Control and Prevention [CDC]. (2010), Morbidity and Mortality Weekly
Report, November 19th, Atlanta, USA.

Chen, S.-H., Wu, V.C.H., Chuang, Y.-C., & Lin, C.-S. (2008). Using oligonucleotide-
functionalized Au nanoparticles to rapidly detect foodborne pathogens on a
piezoelectric biosensor. *Journal of Microbiological Methods 73*, pp. 7–17.

Choi, J.W., Lee, W., Lee, D.B., Park, C.H., Kim, J.S., Jang, Y.H., & Kim, Y. (2007).
Electrochemical Detection of Pathogen Infection Using Cell Chip. *Environ. Monit.
Assess. 129*, pp. 37–42.

Cooper, I.R., Meikle, S.T., Standen, G., Hanlon, G.W., & Santin, M. (2009). The rapid and
specific real-time detection of *Legionella pneumophila* in water samples using Optical
Waveguide Lightmode Spectroscopy. *Journal of Microbiological Methods 78*, pp. 40–
44.

Dawan, S., Kanatharana, P., Wongkittisuksa, B., Limbut, W., Numnuam, A., Limsakul, C., &
Thavarungkul, P. (2011). Label-free capacitive immunosensors for ultra-trace
detection based on the increase of immobilized antibodies on silver nanoparticles.
Analytica Chimica Acta 699 (2), pp. 232-241.

Deobagkar, D.D., Limaye, V., Sinha, S., & Yadava, R.D.S. (2005). Acoustic wave
immunosensing of *Escherichia coli* in water. *Sensors and Actuators B 104*, pp. 85–89.

Diercks, S., Medlin, L.K., & Metfies, K. (2008). Colorimetric detection of the toxic
dinoflagellate Alexandrium minutum using sandwich hybridization in a microtiter
plate assay. *Harmful Algae 7*, pp. 137-145.

Diercks-Horn, S., Metfies, K., Jäckel, S., & Medlin, L.K. (2011). The ALGADEC device: A
semi-automated rRNA biosensor for the detection of toxic algae. *Harmful Algae 10*,
pp. 395-401.

Dinçkaya, E., Kınık, Ö., Sezgintürk, M.K., Altuğ, Ç., & Akkoca, A. (2011). Development of an
impedimetric aflatoxin M1 biosensor based on a DNA probe and gold
nanoparticles. *Biosensors and Bioelectronics 26 (9)*, pp. 3806-3811.

Ding Y., & Mutharasan R. (2010). Highly Sensitive and Rapid Detection of Microcystin-LR in
Source and Finished Water Samples Using Cantilever Sensors. *Environmental
Science and Technology 45(4)*, pp. 1490-1496.

Donaldson, K.A., Kramer, M.F., & Lim, D.V. (2004). A rapid detection method for Vaccinia
virus, the surrogate for smallpox virus. *Biosensors &Bioelectronics 20*, pp. 322-327.

Duplan, V., Frost, E., & Dubowski, J.J. (2011). A photoluminescence-based quantum
semiconductor biosensor for rapid in situ detection of Escherichia coli. *Sensors and
Actuators B*, doi:10.1016/j.snb.2011.07.010.

European Commission [EC], Directive 2006/7/EC of the European Parliament and of the
Council of 2006, February 15th, Concerning the management of bathing water
quality and repealing Directive 76/160/EEC, In: *Official Journal of the European
Communities, L64*, pp. 37-51.

EC, Council Directive 98/83/EC of 1998, November 3st. On the quality of water intended for
human consumption, In: *Official Journal of the European Communities, L330*, pp. 32-54.

EC. Directive 2000/60/CE of 2000, October 23th. Water Frame Directive, In: *Official Journal of
European Commission, L327/1.*

European Environment Agency [EEA]. (2010). The European environment, State and
outlook 2010, Water resources: quantity and flows, In: *Publications Office of the
European Union*, Luxembourg, (ISBN 978 92 9213 162 3).

EEA, Wise TCM hazardous substances. 2011, February 18th. Available from:
    <http://www.eea.europa.eu/themes/water/interactive/tcm-hs>
Erdem, A., Kerman, K., Meric, B., Akarca, U.S., & Ozsoz, M. (2000). Novel hybridization
    indicator methylene blue for the electrochemical detection of short DNA sequences
    related to the hepatitis B virus. *Analytica Chimica Acta 422*, pp. 139–149.
Eurostat, European Commission. (2010). Environmental statistics and accounts in Europe,
    In: *Eurostat Statistical Book.*
Ferrari, B.C., Stoner, K., & Bergquist, P.L. (2006). Applying fluorescence based technology to
    the recovery and isolation of Cryptosporidium and Giardia from industrial
    wastewater streams. *Water Research 40*, pp. 541–548.
Fu, L., Zhang, K., Li, S., Wang, Y., Huang, T.-S., Zhang, A., & Cheng, Z.-Y. (2010). In situ
    real-time detection of E. coli in water using antibody-coated magnetostrictive
    microcantilever. *Sensors and Actuators B 150*, pp. 220–225.
Gau J.-J., Lan E.H., Dunn B., Ho C.-M., & Woo J.C.S. (2001). A MEMS based amperometric
    detector for *E. Coli* bacteria using self-assembled monolayers. *Biosensors &
    Bioelectronics 16*, pp. 745–755.
Geng, P., Zhang, X., Teng, Y., Fu, Y., Xu, L., Xu, M., Jin, L., & Zhang, W. (2011). DNA
    sequence-specific electrochemical biosensor based on alginic acid-coated cobalt
    magnetic beads for the detection of E. coli. *Biosensors and Bioelectronics 26*, pp. 3325–
    3330.
Gilgen, M., Germann, D., Lüthy, J., & Hübner, Ph. (1997). Three-step isolation method for
    sensitive detection of enterovirus, rotavirus, hepatitis A virus, and small round
    structured viruses in water samples. *International Journal of Food Microbiology 37*, pp.
    189–199.
Girones, R., Ferrus, M.A., Alonso, J.L., Rodriguez-Manzano, J., Calgua, B., Corrêa, A.,
    Hundesa, A., Carratala, A., & Bofill-Mas, S. (2010). Molecular detection of
    pathogens in water - The pros and cons of molecular techniques. *Water Research 44*,
    pp. 4325-4339.
Gregora J., & Marsálek B. (2005). A Simple In Vivo Fluorescence Method for the Selective
    Detection and Quantification of Freshwater Cyanobacteria and Eukaryotic Algae.
    Acta hydrochim. hydrobiol. 33, pp. 142−148.
Guntupalli, R., Lakshmanan, R.S., Hu, J., Huang, T.S., Barbaree, J.M., Vodyanoy, V., & Chin,
    B.A. (2007). Rapid and sensitive magnetoelastic biosensors for the detection of
    Salmonella typhimurium in a mixed microbial population. *Journal of Microbiological
    Methods 70*, pp. 112–118.
Guven, B., Basaran-Akgul, N., Temur, E., Tamer, U., & Boyac, I.H. (2011). SERS-based
    sandwich immunoassay using antibody coated magnetic nanoparticles for
    Escherichia coli enumeration. *Analyst 136*, pp. 740–748.
Haughey, S.A., Campbell, K., Yakes, B.J., Prezioso, S.M., DeGrasse, S.L., Kawatsu, K., &
    Elliott C.T. (2011). Comparison of biosensor platforms for surface plasmon
    resonance based detection of paralytic shellfish toxins. *Talanta 85 (1)*, pp. 519-526.
Health Protection Agency. (2005). Enumeration of coliforms and Escherichia coli by Idexx
    (colilert 18) Quanti-trayTM. National Standard Method W 18.
    <http://www.hpastandardmethods.org.uk/pdf_sops.asp.>

Henkel, J.H., Aberle, S.W., Kundi, M., & Popow-Kraupp, T. (1997). Improved detection of respiratory syncytial virus in nasal aspirates by seminested RT-PCR. *J. Med. Virol. 53*, pp. 366-371.

Hong, P.-Y., Hwang, C., Ling, F., Andersen, G.L., LeChevallier, M.W., Liu, W.-T. (2010). Pyrosequencing analysis of bacterial biofilm communities in water meters of a drinking water distribution system. *Applied and Environmental Microbiology 76*, pp. 5631-5635.

Hu C., Gan N., He Z., & Song L. (2008). A novel chemiluminescent immunoassay for microcystin (MC) detection based on gold nanoparticles label and its application to MC analysis in aquatic environmental samples. *Intern. J. Environ. Anal. Chem. 88*, pp. 267–277.

Huang, S., Pang, P., Xiao, X., He, L., Cai, Q., & Grimes, C.A. (2008). A wireless, remote-query sensor for real-time detection of *Escherichia coli* O157:H7 concentrations. *Sensors and Actuators B 131*, pp. 489–495.

Huang, Y., Dong, X., Liu, Y., Lic, L.-J., & Chen, P. (2011). Graphene-based biosensors for detection of bacteria and their metabolic activities. *Journal of Materials Chemistry 21*, pp. 12358-12362.

Hunter P., Lee J., Nichols G., Rutter M., Surman S., Weldon L., Biegon D., Fazakerley T., Drobnewski F., & Morrell P. (2001). Fate of *Mycobacterium avium Complex* in Drinking Water Treatment and Distribution Systems. Research report DWI 70/2/122.

Ilic, B., Yang, Y., & Craighead, H. G. (2004). Virus detection using nanoelectromechanical devices. *Appl. Phys. Lett.85*, pp. 2604-2606.

International Organization for Standardization [ISO]. (2005). ISO 8199: 2005, Water quality – General Guidance on the enumeration of microorganisms by culture.

Jing G., Polaczyk A., Oerther D.B., & Papautsky I. (2007). Development of a microfluidic biosensor for detection of environmental mycobacteria. *Sensors and Actuators B 123*, pp. 614–621.

Karsunke X.Y.Z., Niessner R., & Seidel, M. (2009). Development of a multichannel flow-through chemiluminescence microarray chip for parallel calibration and detection of pathogenic bacteria. *Anal Bioanal Chem 395*, pp. 1623–1630.

Kattke, M.D., Gao, E.J., Sapsford, K.E., Stephenson, L.D., & Kumar, A. (2011). FRET-Based Quantum Dot Immunoassay for Rapid and Sensitive Detection of Aspergillus amstelodami. *Sensors 11*, pp. 6396-6410.

Kormas, K., Neofitou, C., Pachiadaki, M., & Koufostathi, E. (2010). Changes of the bacterial assemblages throughout an urban drinking water distribution system. *Environmental Monitoring Assess. 165*, pp. 27-38.

Kuznetsov, Y.G., Daijogo, S., Zhou, J., Semler, B.L., & McPherson, A. (2005). Atomic force microscopy analysis of icosahedral virus RNA. *J. Mol. Biol. 347*, pp. 41-52.

Kwon, H.-J., Dean, Z.S., Angus S.V., & Yoon, J.-Y. (2010). Lab-on-a-chip for Field *Escherichia coli* Assays: Long-term Stability of Reagents and Automatic Sampling System. *JALA - Journal of the Association for Laboratory Automation 15*, pp. 216-223.

Lassiter, K., Wang, R., Lin, J., Lum, J., Srinivasan, B., Lin, L., Lu, H., Hargis, B., Bottje, W., Tung, S., Berghman, L., & Li, Y. (2009). Comparison study of an impedance biosensor and rRT-PCR for detection of avian influenza H5N2 from infected chickens. *Program Book of PSA 2009 Annual Meeting*, Raleigh, NC, July, 2009.

Leonard, P., Hearty, S., Brennan, J., Dunne, L., Quinn, J., Chakraborty, T., & O'Kennedy, R. (2003). Advances in biosensors for detection of pathogens in food and water. *Enzyme and Microbial Technology 32*, pp. 3–13.

Li, K., Lai, Y., Zhang, W., & Jin, L. (2011A). Fe$_2$O$_3$@Au core/shell nanoparticle-based electrochemical DNA biosensor for *Escherichia coli* detection. *Talanta 84*, pp. 607–613.

Li, S., Chen, J., Cao, H., Yao, D., & Liu, D. (2011B). Amperometric biosensor for aflatoxin B1 based on aflatoxin-oxidase immobilized on multiwalled carbon nanotubes. *Food Control 22 (1)*, pp. 43-49.

Li, X.X., Cao, C., Han, S.J., & Sim, S.J. (2009). Detection of pathogen based on the catalytic growth of gold nanocrystals. *Water Research 43*, pp. 1425–1431.

Lin, H.-Y., Tsao, Y.-C., Tsai, W.-H., Yang, Y.-W., Yan, T.-R., & Sheu, B.-C. (2007). Development and application of side-polished fiber immunosensor based on surface plasmon resonance for the detection of *Legionella pneumophila* with halogens light and 850 nm-LED. *Sensors and Actuators A 138*, pp. 299–305.

Lindner, P., Molz, R., Yacoub-George, E., & Wolf, H. (2009). Rapid chemiluminescence biosensing of microcystin-LR. *Analytica Chimica Acta 636 (2)*, pp. 218-223.

Liu Y., & Li Y. (2002). Detection of *Escherichia coli* O157:H7 using immunomagnetic separation and absorbance measurement. *Journal of Microbiological Methods 51*, pp. 369– 377.

Liu, X., Cheng, Z., Fan, H., Ai, S., & Han R. (2011). Electrochemical detection of avian influenza virus H5N1 gene sequence using a DNA aptamer immobilized onto a hybrid nanomaterial-modified electrode. *Electrochimica Acta 56 (18)*, pp. 6266-6270.

Loge, F.J., Thompson, D.E., Call, D.R. (2002). PCR Detection of Specific Pathogens in Water: A Risk-Based Analysis. *Environ. Sci. Technol. 36*, pp. 2754-2759.

Long F., He M., Zhu A.N., & Shi H.C. (2009). Portable optical immunosensor for highly sensitive detection of microcystin-LR in water samples. *Biosensors and Bioelectronics 24*, pp. 2346-2351.

Long, F., He, M., Zhu, A., Song, B., Sheng, J., & Shi, H. (2010). Compact quantitative optic fiber-based immunoarray biosensor for rapid detection of small analytes. *Biosensors and Bioelectronics 26 (1)*, pp. 16-22.

Loypraserta, S., Thavarungkul, P., Asawatreratanakul, P., Wongkittisuksa, B., Limsakul, C., & Kanatharana, P. (2008). Label-free capacitive immunosensor for microcystin-LR using self-assembled thiourea monolayer incorporated with Ag nanoparticles on gold electrode. *Biosensors and Bioelectronics 24 (1)*, pp. 78-86.

Lum, J., & Li, Y. (2010). Rapid detection of avian influenza H5N1 virus using impedance measurement of immuno-reaction coupled with RBC amplification. Master Thesis, Pub. N.º 1485925, University of Arkansas.

Luo, Y., Nartker, S., Miller, H., Hochhalter, D., Wieroder, M., Wieroder, S., Setterington, E., Drzal, L.T., & Alocilja, E.C. (2010). Surface functionalization of electrospun nanofibers for detecting E. coli O157:H7 and BVDV cells in a direct-charge transfer biosensor. *Biosensors and Bioelectronics 26*, pp. 1612–1617.

Metfies, K., Huljic, S., Lange, M., & Medlin, L.K. (2005). Electrochemical detection of the toxic dinoflagellate Alexandrium ostenfeldii with a DNA-biosensor. *Biosensors and Bioelectronics 20*, pp. 1349–1357.

Miranda-Castro, R., de-los-Santos-Álvarez, N., Lobo-Castañón, M.J., Miranda-Ordieres, A.J., & Tuñón-Blanco, P. (2009). PCR-coupled electrochemical sensing of *Legionella pneumophila*. *Biosensors and Bioelectronics 24*, pp. 2390–2396.

Mons, C., Dumètre, A., Gosselin, S., Galliot, C., & Moulin, L. (2009). Monitoring of Cryptosporidium and Giardia river contamination in Paris area. *Water Research 43*, pp. 211–217.

Nayak, M., Kotian, A., Marathe, S., & Chakravortty, D. (2009). Detection of microorganisms using biosensors- A smarter way towards detection techniques. *Biosensors and bioelectronics 25*, pp. 661-667.

O'Shea, M. K., Ryan, M.A.K., Hawksworth, A.W., Alsip, B.J., & Gray, G.C. (2005). Symptomatic Respiratory Syncytial Virus Infection in Previously Healthy Young Adults Living in a Crowded Military Environment. *Clin. Infect. Dis. 41*, pp. 311-317.

Orozco, J., & Medlin, L.; 2011b. Electrochemical performance of a DNA-based sensor device for detecting toxic algae. *Sensors and Actuactors B 153*, pp. 71–77.

Orozco, J., Baudart, J., & Medlin, L.K. (2011). Evaluation of probe orientation and effect of the digoxigenin-enzymatic label in a sandwich hybridization format to develop toxic algae biosensors. *Harmful Algae 10 (5)*, pp. 489-494.

Park, E.J., Lee, J.-Y., Kim, J.H., Lee, C.J., Kim, H.S., & Min, N.K. (2010). Investigation of plasma-functionalized multiwalled carbon nanotube film and its application of DNA sensor for *Legionella pneumophila* detection. *Talanta 82*, pp. 904–911.

Park, S., Kim, H., Paek, S.-H., Hong, J.W., & Kim, Y.-K. (2008). Enzyme-linked immuno-strip biosensor to detect *Escherichia coli* O157:H7. *Ultramicroscopy 108*, pp. 1348– 1351.

Paterson R.R.M., Kelley J., & Gallagher M. (1997). Natural occurrence of aflatoxins and Aspergillus flavus (LINK) in water. *Lett. Appl. Microbiol. 25*, pp. 435–436.

Poitelon, J.-B., Joyeux, M., Welté, B., Duguet, J.-P., Prestel, E., DuBow, M. (2010). Variations of bacterial 16S rRNA phylotypes prior to and after chlorination for drinking water production from two surface water treatment plants. *J. Ind. Microbiol. Biotechnol. 37*, pp. 117-128.

Pyo, D., & Jin, J. (2007). Production and Degradation of Cyanobacterial Toxin in Water Reservoir, Lake Soyang. *Bull. Korean Chem. Soc. 28 (5)*, pp. 800-804.

Pyo, D., Choi, J., Hong, J., & Oo, H.H. (2006). Rapid Analytical Detection of Microcystins Using Gold Colloidal Immunochromatographic Strip. *Journal of Immunoassay & Immunochemistry 27*, pp. 291–302.

Pyo, D., Huang, Y., Kim, Y., & Hahn, J.H. (2005). Ultra-Sensitive Analysis of Microcystin-LR Using Microchip Based Detection System. *Bull. Korean Chem. Soc. 2005 26 (6)*, pp. 939-942.

Queirós, R.B., Silva, S.O., Noronha, J.P., Frazão, O., Jorge, P., Aguilar, G., Marques, P.V.S., & Sales, M.G.F. (2011). Microcystin-LR detection in water by the Fabry–Pérot interferometer using an optical fibre coated with a sol–gel imprinted sensing membrane. *Biosensors and Bioelectronics 26*, pp. 3932–3937.

Rasooly, A., & Herold, K.E. (2006) Biosensors for the Analysis of Food- and Waterborne Pathogens and Their Toxins. *Journal of AOAC International 89*, pp. 873-883.

Revetta, R., Pemberton, A., Lamendella, R., Iker, B., Santo Domingo, J.W. (2010). Identification of bacterial populations in drinking water using 16S rRNA-based sequence analyses. *Water Research 44*, pp. 1353-1360.

Russell R., Paterson M., & Lima N. (2005). Fungal Contamination of Drinking Water, In: *Water Encyclopedia*, John Wiley and Sons, Inc..

Russell, R., & Paterson, M. (2007). Zearalenone production and growth in drinking water inoculated with Fusarium graminearum. *Mycological Progress 6 (2)*, pp. 109-113.

Russell, R., Paterson, M., & Lima, N. (2010) The Weaponisation of Mycotoxins (Chapter 21). In: *Mycotoxins in Food, Feed and Bioweapons*. Springer-Verlag, Berlin, Heidelberg.

Sapsford, K.E., Ngundi, M.M., Moore, M.H., Lassman, M.E., Shriver-Lake, L.C., Taitt, C.R., & Ligler, F.S. (2006). Rapid detection of foodborne contaminants using an Array Biosensor. *Sensors and Actuators B 113*, pp. 599–607.

Shanmukh, S., Jones, L., Driskell, J., Zhao, Y., Dluhy, R., & Tripp, R.A. (2006). Rapid and Sensitive Detection of Respiratory Virus Molecular Signatures Using a Silver Nanorod Array SERS Substrate. *Nanoletters 6 (11)*, pp. 2630-2636.

Sheng, J., He, M., Yu, S., Shi, H., & Qian, Y. (2007). Microcystin-LR detection based on indirect competitive enzyme-linked immunosorbent assay. *Front. Environ. Sci. Engin. China 1(3)*, pp. 329–333.

Sigee, D. C. (2005). *Freshwater Microbiology*. John Wiley and Sons, LTD, Manchester, UK.

Simpson, J.M., & Lim, D.V. (2005). Rapid PCR confirmation of *E. coli* O157:H7 after evanescent wave fiber optic biosensor detection. *Biosensors and Bioelectronics 21*, pp. 881-887.

Smith, H.V., & Nichols, R.A.B. (2010). Cryptosporidium: Detection in water and food. *Experimental Parasitology 124*, pp. 61–79.

Song, S., Wang, L., Li, J., Zhao, J., & Fan, C. (2008). Aptamer-based biosensors. *Trends in Analytical Chemistry 27*, pp.108-117.

Soule, H., Genoulaz, O., Gratacap-Cavallier, B., Chevallier, P., Liu, J.-L., & Seigneurin J.-M. (2000). Ultrafiltration ans reverse transcription-polymerase chain reaction: an efficient process for polivirus, rotavirus and hepatitis A virus detection in water. *Water Research 34 (3)*, pp. 1063-1067.

Spichiger-Keller U.E. (1998). Chemical and Biochemical Sensors (Chapter 2), In: *Chemical Sensors and Biosensors for Medical and Biological Applications*, Weinheim, WCH-Wiley.

Su, L., Jia, W., Hou, C., & Lei, Y. (2011). Microbial biosensors: a review. *Biosensors and bioelectronics 26*, pp. 1788-1799.

Sun, H., Choy, T.S., Zhu, D.R., Yam, W.C., & Fung, Y.S. (2009). Nano-silver-modified PQC/DNA biosensor for detecting *E. coli* in environmental water. *Biosensors and Bioelectronics 24*, pp. 1405–1410.

Sungkanak, U., Sappat, A., Wisitsoraat, A., Promptmas, C., & Tuantranont, A. (2010). Ultrasensitive detection of *Vibrio cholerae* O1 using microcantilever-based biosensor with dynamic force microscopy. *Biosensors and Bioelectronics 26*, pp. 784–789.

Tang, H., Zhang, W., Geng, P., Wang, Q., Jin, L., Wu, Z., & Lou, M. (2006). A new amperometric method for rapid detection of *Escherichia coli* density using a self-assembled monolayer-based bienzyme biosensor. *Analytica Chimica Acta 562*, pp. 190–196.

Tavares, T., Cardoso, D., & Diederichsen de Brito, W. (2005). Vírus entéricos veiculados por água: aspectos microbiológicos e de controle de qualidade da água. *Revisão 34*, pp. 85-104.

Thiruppathiraja, C., Kamatchiammal, S., Adaikkappan, P., Santhosh, D.J., & Alagar, M. (2011). Specific detection of *Mycobacterium sp.* genomic DNA using dual labeled gold nanoparticle based electrochemical biosensor. *Analytical Biochemistry 417*, pp. 73–79.

Tobler N.E., Pfunder M., Herzog K., Frey J.E., & Altwegg M. (2006). Rapid detection and species identification of *Mycobacterium spp.* using real-time PCR and DNA-Microarray. *Journal of Microbiological Methods 66*, pp. 116– 124.

Toze S. (1999). PCR and the Detection of Microbial pathogens in Water and Wastewater. *Water. Research 33*, pp. 3545-3556.

Turner, A.P.F., & Piletsky S. (2005). Biosensors and biommimetic sensors for the detection of drugs, toxins and biological agents., In: *NATO Security through Science Series*, Springer Ed., Volume 1, pp. 261-272.

Vo-Dinh, T. (2007). Biosensors and biochips, In: *Biomolecular sensing, processing and analysis*, Volume I, pp. 3-20.

Vörös, J., Ramsden, J.J., Csucs, G., Szendrö, I., Textor M., & Spencer, N.D. (2002). Optical Grating Coupler Biosensors. *Biomaterials 23 (17)*, pp. 3699-3710.

Wang L., Chen W., Xu D., Shim, B.S., Zhu Y., Sun F., Liu L., Peng C., Jin Z., Xu C., & Kotov N.A. (2009). Simple, rapid, sensitive, and versatile SWNT- Paper sensor for environmental toxin detection competitive ELISA. *Nano Letters 9(12)*, pp. 4147-4152.

Wang, L., Liu, Q., Hu, Z., Zhang, Y., Wu, C., Yang, M., & Wang, P. (2009). A novel electrochemical biosensor based on dynamic polymerase-extending hybridization for E. coli O157:H7 DNA detection. *Talanta 78*, pp. 647–652.

World Health Organization [WHO] and WASH Inter Agency Group. (2007) Questions & Answers on potential transmission of avian influenza (H5N1) through water, In: *Sanitation and Hygiene and ways to reduce the risks to human health.*

WHO. (1999). Toxic Cyanobacteria in Water: A guide to their public health consequences, monitoring and management. E & FN Spon, London.

WHO. (2006). Review of latest available evidence on potential transmission of avian influenza (H5N1) through water and sewage and ways to reduce the risks to human health, In: *Water, Sanitation and Health Public Health and Environment*, Geneva.

WHO. (2011). Global Alert and Response Programme. Cholera outbreaks: Cholera outbreaks in the Democratic Republic of Congo (DRC) and the Republic of Congo, July 22nd, 2011.

Wildeboer, D., Amirat, L., Price, R.G., & Abuknesha, R.A. (2010). Rapid detection of *Escherichia coli* in water using a hand-held fluorescence detector. *Water Research 44*, pp. 2621–2628.

Wolter, A., Niessner, R., & Seidel, M. (2008). Detection of *Escherichia coli* O157:H7, *Salmonella typhimurium*, and *Legionella pneumophila* in Water Using a Flow-Through Chemiluminescence Microarray Readout System. *Anal. Chem. 80*, pp. 5854–5863.

Xia, Y., Deng, J., & Jiang, L. (2010). Simple and highly sensitive detection of hepatotoxin microcystin-LR via colorimetric variation based on polydiacetylene vesicles. *Sensors and Actuators B: Chemical145 (2)*, pp. 713-719.

Xia, Y., Zhanga, J., & Jiang, L. (2011). A novel dendritic surfactant for enhanced microcystin-LR detection by double amplification in a quartz crystal microbalance biosensor. *Colloids and Surfaces B: Biointerfaces 86 (1)*, pp. 81-86.

Xu, J., Suarez, D., & Gottfried, D. (2007). Detection of Avian Influenza Virus Using An Interferometric Biosensor. *Analytical and Bioanalytical Chemistry 389*, pp. 1193-1199.

Xue, X., Pan, J., Xie, H., Wang, J., & Zhang, S. (2009). Fluorescence detection of total count of *Escherichia coli* and *Staphylococcus aureus* on water-soluble CdSe quantum dots coupled with bacteria. *Talanta 77*, pp. 1808–1813.

Yakes, B.J., Prezioso, S., Haughey, S.A., Campbell, K., Elliott, C.T., & DeGrasse, S.L. (2011) An improved immunoassay for detection of saxitoxin by surface plasmon resonance biosensors. *Sensors and Actuators B: Chemical 156 (2)*, pp. 805-811.

Yáñez M.A, Valor C., & Catalán V. (2006). A simple and cost-effective method for the quantification of total coliforms and *Escherichia coli* in potable water. *Journal of Microbiological Methods 65*, pp. 608–611.

Ye, Y.K., Zhao, J.H., Yan, F., Zhu, Y.L., & Ju, H.X. (2003). Electrochemical behavior and detection of hepatitis B virus DNA PCR production at gold electrode. *Biosensors and Bioelectronics 18*, pp. 1501-1508.

Yoo, S.K., Lee, J.H., Yun, S.-S., Gu, M.B., & Lee, J.H. (2007). Fabrication of a bio-MEMS based cell-chip for toxicity monitoring. *Biosensors and Bioelectronics 22*, pp. 1586–1592.

Yoon, J.-Y., Han, J.-H., Choi, C.Y., Bui, M., & Sinclair, R.G. (2009). Real-Time Detection of Escherichia coli in Water Pipe Using a Microfluidic Device with One-Step Latex Immunoagglutination Assay. *Transactions of the ASABE, 52(3)*, pp. 1031-1039.

Yu J., Liu Z., Liu Q., Yuen K.T., Mak A.F.T., Yang M., & Leung P. (2009). A polyethylene glycol (PEG) microfluidic chip with nanostructures for bacteria rapid patterning and detection. *Sensors and Actuators A 154*, pp. 288–294.

Yu, C.Y., Ang, G.Y., Chua, A.L., Tan, E.H., Lee, S.Y., Falero-Diaz, G., Otero, O., Rodríguez, I., Reyes, F., Acosta, A., Sarmiento, M.E., Ghosh, S., Ramamurthy, T., Yean, C.Y., Lalitha, P., & Ravichandran, M. (2011). Dry-reagent gold nanoparticle-based lateral flow biosensor for the simultaneous detection of *Vibrio cholerae* serogroups O1 and O139. *Journal of Microbiological Methods 86*, pp. 277–282.

Yuan, J., Deng, D., Lauren, D.R., Aguilar, M.-I., & Wu, Y. (2009). Surface plasmon resonance biosensor for the detection of ochratoxin A in cereals and beverages. *Analytica Chimica Acta 656 (1-2)*, pp. 63-71.

Zhang, F., Yang, S.H., Kang, T.Y., Cha, G.S., Nam, H., & Meyerhoff, M.E. (2007). A rapid competitive binding nonseparation electrochemical enzyme immunoassay (NEEIA) test strip for microcystin-LR (MCLR) determination. *Biosensors and Bioelectronics 22 (7)*, pp. 1419-1425.

Zheng, Y.Z., Hyatt, A., Wang, L.F., Eaton, B.T., Greenfield, P.F., & Reid, S.J. (1999). Quantification of recombinant core-like particles of bluetongue virus using immunosorbent electron microscopy. *Virol. Methods 80*, pp. 1-9.

Zhou, G., Wen, S., Liu, Y., Li, R., Zhong X., Feng, L., Wang, L., & Cao, B. (2011). Development of a DNA microarray for detection and identification of *Legionella pneumophila* and ten other pathogens in drinking water. *International Journal of Food Microbiology 145*, pp. 293–300.

8

Posidonia oceanica and *Zostera marina* as Potential Biomarkers of Heavy Metal Contamination in Coastal Systems

Lila Ferrat[1] et al.*
[1]*University of Corsica, Sciences for Environment,*
France

1. Introduction

In the early 1960s recognition of the adverse effects of environmental contamination due to industrial, pesticide, and agricultural pollution led to the emergence of the field of ecotoxicology (Ramade, 1992). Today, marine estuary and inshore ecosystems continue to be negatively impacted by environmental contamination (Short & Wyllie-Echeverria 1996; Orth et al., 2006; Osborn & Datta, 2006). In order to reduce these negative impacts, bio-surveillance programs are needed to monitor environmental conditions so that changes in ecosystem processes, structure, and the physiological condition of species can be assessed (Blandin, 1986; Tett et al., 2007). An important characteristic of these programs is that indicator species must be capable of rapidly detecting significant changes in the ecosystem so that the cause of deterioration can be addressed early (e.g. Hemminga & Duarte, 2000).

Mussels (Goldberg et al., 1983) and fish (Reichert et al., 1998; Stephensen et al., 2000) are frequently used as indicators of chemical contamination in long-term environmental monitoring programs. However, these programs can be deficient because they only provide information about water column contamination, and these organisms can have limited ranges and often must be introduced to a site as part of the monitoring program. To offset these deficiencies widely distributed indicator organisms in coastal systems that have the capacity to provide contamination information from both water column and sediment environments are needed. Consequently, there is increasing interest in the use of marine macrophytes because they grow in most coastal and estuarine systems (see Green & Short, 2003). These rooted vascular plants interact with the chemical properties of the water column and surface sediment environments within site-specific and basin-wide locations (Brix, 1997; Brix et al., 1983; den Hartog, 1970; Lange and Lambert, 1994; Rainbow and

---

*Sandy Wyllie-Echeverria[2], G. Cates Rex[3], Christine Pergent-Martini[1], Gérard Pergent[1], Jiping Zou[3], Michèle Romeo[4], Vanina Pasqualini[1] and Catherine Fernandez[5]
[1]*University of Corsica, Sciences for Environment, France*
[2]*Friday Harbor Laboratories, University of Washington USA*
[3]*Brigham Young University, Department of Biology, USA*
[4]*University of Nice Sophia-Antipolis, ROSE, France*
[5]*University of Aix-Marseille, IMEP, France*

Phillips, 1993). For this study the focus was on the seagrasses *Posidonia oceanica* (L.) Delile (Posidoniaceae) and *Zostera marina* L. (Zosteraceae). These were chosen because they are the dominant species in the regions of our inquiry which were, respectively, the Mediterranean Sea (Lipkin et al., 2003; Procaccini et al., 2003) and the Pacific Northwest (Wyllie-Echeverria & Ackerman, 2003). Both species can form vast meadows across the intertidal-subtidal gradient in their respective ecosystems (Molinier & Picard, 1952; Phillips, 1984).

1.1 *P. oceanica* and *Z. marina* as indicators of environmental quality

The potential for these species to provide an early warning of deteriorating environmental quality has been noted for *P. oceanica* and *Z. marina* where both species were found useful at detecting environmental deterioration within local and basin-wide locations (Augier, 1985; Dennison et al., 1993; Pergent, 1991; Pergent-Martini et al., 1999). For example, *P. oceanica* accumulates certain metal pollutants, notably mercury (Pergent-Martini, 1998), which is one of the most abundant marine pollutants. Within the Mediterranean Sea elevated mercury levels have been reported in certain regions (Maserti et al., 1991), and correlations have been drawn between mercury levels in plant tissue and the concentrations of mercury in the water column (Pergent-Martini, 1998). In laboratory studies Lyngby & Brix (1984) and Brix & Lyngby (1984) demonstrated that *Z. marina* can accumulate heavy metals in concentrations above natural levels, and that these concentrations inhibited growth. In addition, based on extensive sampling along the coastline of Limfjord, Denmark, these authors noted that *Z. marina* could be used to monitor heavy metal contamination. Also, a related species *Z. capricorni* has provided valuable information in monitoring iron, aluminium, zinc, chromium, and copper contamination (Prange & Dennison, 2000).

Indicator species that provide an early warning of ecosystem change will likely be those that reveal first order changes in organism function. Molecular, biochemical, and/or cellular changes triggered by pollutants are measurable in biological mediums such as cells, tissues, and/or cellular fluids (McCarthy & Shugart, 1990). For example, oxidation is known to be a significant factor in stress-related organismal weakening, and antioxidant molecules have been used to evaluate organism health (Chen et al., 2007). One group of antioxidant molecules are the widely studied phenolic compounds (Ferrat et al., 2003a) which are known to be induced by reactive oxygen species (Rice-Evans et al., 1995; Vangronsveld et al., 1997).

1.2 Physiological and ecological roles of phenolics and volatile compounds

Phenolic compounds produced via the Shikimic Acid Pathway, and volatiles produced via the Mevalonate Pathway, are known to be important to plant health and survival (Cates, 1996; Fierer et al., 2001; Hartman, 2007; Phillips, 1992; Schimel et al., 1996). They are found in terrestrial higher plants, most notably angiosperms (Goodwin & Mercer, 1983; Hadacek, 2002), some seagrasses (Verges et al., 2007; Zapata and McMillan, 1979), and have a wide range of chemical structures and activities (Hadacek, 2002; Hartman, 2007). Phenolic and volatile compounds contribute significantly to the antioxidant activity of plants, have the capacity to bind heavy metals (Emmons et al., 1999), and are an important mechanism in protecting plants against stress (Swain, 1977). Volatile compounds (e.g. monoterpenes, sesquiterpenes) have been found to serve as energy sources in plants (Croteau & Sood,

1985), are important in the defensive system of higher plants (Cates, 1996; Langenheim, 1994), and influence ecosystem processes such as nutrient cycling (Horner et al., 1988; White, 1986). The production of phenolics and volatiles is under genetic control (Croteau & Gershenzon, 1994; Hartman, 2007), but their qualitative and quantitative production is affected by various environmental factors (Bryant et al., 1983; Gershenzon, 1984; Hartman, 2007; Macheix, 1996; Quackenbush et al., 1986; Ragan & Glombitza, 1986). However, as with other seagrasses, only a very limited number of studies deal with the role of phenolic and volatile compounds from *Posidonia oceanica* (Heglmeier & Zidorn, 2010) and *Zostera marina* (Short & Willie Echeverria, 1996). Only were investigated the impacts of interspecific competition (Dumay et al., 2004), nutrient variation, diseases (Vergeer & Develi, 1997) and grazing (Cannac et al., 2006), or general anthropization of water masses (Short & Wyllie-Echeverria, 1996, Agostini et al., 1998).

1.3 Objective

The objective of this study was to determine if *P. oceanica* and *Z. marina* might be reliable candidates as bio-surveillance organisms with regard to heavy metal pollution. We choose to consider different environmental conditions and to monitor physiological changes through two different seasons. Our assumption was that heavy metal contamination would adversely impact adult *P. oceanica* and *Z. marina* plants, and that plant response to these impacts could be assessed by differences in phenolic and volatile compound content of tissue from impacted and non-impacted sites.

We assessed differences in heavy metal content of plant tissues from sites with documented heavy metal pollution versus controls with no sources of heavy metal pollution. Then, we tested the hypothesis that the presence of identified contaminants could induce a bio-indicator response in these seagrass species. To do this we measured changes in total phenolic content in the leaf and sheath tissue of *P. oceanica*, and total phenolic and volatile compound content in above-and below-ground tissue of *Z. marina*.

2. Materials and methods

2.1 Site location and sample collection

In June 2000 and January 2001, 30 adult shoots of *P. oceanica* were collected by SCUBA at ~10 m in the sub-tidal region at two sites located in the northwestern Mediterranean Sea. The Bay of Bonifacio, a control site, is a pristine area relatively free of industrial pollution located in the south of Corsica (Tonnara - France; 41.4000 N; 9.0830 E; Capiomont et al., 2000). The Bay of Rosignano site south of Livorno (Italy; 43.4000 N; 10.4166 E) is a polluted site. At this site, a chlor-alkali plant has discharged industrial wastes rich in mercury since 1920 (130 kg per year; Ferrara et al., 1989). Water temperature ranged from 18°C in June 2000 to 14°C in January 2001 at all sites but salinity was relatively constant at 38.5 PSU within the study zone (i.e., 10 m depth contour; Villefranche sur Mer Observatory and Di Martino, personal communication).

For *P. oceanica*, foliage leaf and sheath tissue was analyzed for mercury and phenolic content. Tissue was obtained by separating the foliage leaf and sheath tissue from the roots and rhizomes following the procedure of Giraud (1977); root and rhizome tissue was discarded. The chlorophyllous foliage leaves were then separated from non-chlorophyllous sheaths that are located at the leaf base. Foliage leaves from three adult shoots were dissected according to Giraud (1977) and combined to form one sample. Sheaths from the same three shoots were combined to form each sheath sample. After epiphytes were removed from leaf and sheath samples using a glass slide, each sample was rinsed with ultra-pure water and frozen (-20°C) until analysis. To determine mercury and phenolic content, we extracted 0.5 g dry wt. of each tissue sample (n=10).

During maximum low tide, Z. marina adult shoots were hand-collected from the lower intertidal region of two sites in Northern Puget Sound, Washington, USA in April and June 2000. The site located near Anacortes, WA (48.4263 N; 122.5897 W) was documented as having heavy metal pollution (http://www.ecy.wa.gov/programs/wq/permits/permit_pdfs/dakota/factsheet.pdf), and the other was a pristine location with no industrial activity on the southeast side of Shaw Island (48.33942 N; 122.55448 W) that served as the control site (Wyllie-Echeverria & Ackerman, 2003). Water temperature ranged between 9°C in April to12°C in June 2000 at all sites, and salinity was relatively constant at 30 PSU during this time period (Wyllie-Echeverria, unpublished data).

Three samples were collected from each site, and each sample consisted of at least 0.5 g dry wt (Cuny et al., 1995) of eight to ten sterile (non-reproductive) shoots which were separated into above- and below-ground parts. Above-ground tissue consisted of the foliage leaf (i.e. basal leaf sheath and distal leaf blade; Kuo & den Hartog, 2006) excised from the rhizome at the node primordia (Tomlinson, 1974). The remaining rhizome and associated nodes and roots formed the below-ground sample. Epiphytes were scraped from the above-ground tissue and sediment was rinsed from the roots and rhizomes (Brackup & Capone, 1985). Each above- and below-ground sample was placed in labelled bags, kept moist and cool in a refrigerator, and shipped overnight to the Chemical Ecology Laboratory at Brigham Young University. Three replicate samples of above-ground tissue from each site, and three replicates of below-ground tissue from each site, were frozen at –80°C until extracted for heavy metals, phenolics or volatiles. Samples were stored at –80°C to preserve the volatile compounds in the tissues.

2.2 Qualitative and quantitative analysis of plant tissues for heavy metal content

Foliage leaves and sheaths of P. oceanica and above- and below-ground tissues of Z. marina were analyzed qualitatively and quantitatively for heavy metals. For P. oceanica, only mercury content, which is the predominant heavy metal pollutant at the Rosignano site (Lafabrie et al., 2007), was analyzed. Three individual shoots (three foliage leaves and three sheaths) that had been separately freeze dried were ground to a powder, and an aliquot of 0.05 g dry wt was digested. Digestion was performed in a 100-ml Teflon® advanced composite vessel reactor with 5 ml HNO_3 and 1 ml H_2O_2 (30%). Microwave digestion (Mars 5, CEM Chemistry, Engineering and Microwave, Matthews, NC, USA) was carried out using a temperature ramp of 8 min up to 200°C followed by a heating plateau of 20 min at 200°C. After digestion, the samples were increased to 25 ml with ultra-pure water and then filtered. Total mercury was determined using a flameless atomic absorption spectrophotometer flow injection (Perkin-Elmer System 100; Norwalk, CT, USA). The procedure consisted of reduction with 1.1% tin chloride ($SnCl_2$, $2H_2O$) in 3% HCl and 0.5% hydroxylammonium chloride (NH_2OH, HCl). A standard addition method for total mercury was used to calibrate the protocol. The analytic procedure was verified using a moss as the certified reference material (*Lagarosiphon major*, Certified Reference Material 60, Community Bureau of Reference, Commission of the European Community, Brussels, Belgium). Data are expressed as ng per g dry wt.

For *Z. marina*, heavy metal content was analyzed using the EPA Method 3052 Procedure. All elements were wet-ashed to prevent loss of elements and reduce the potential of confounding data due to silica content. Above- and below-ground tissue (0.5 g dry wt) was placed in a 50 ml folin tube, and 5 ml concentrated nitric acid was added. Samples sat overnight, and then were placed on a block digester at 200°C for 5-10 minutes. Tubes were removed, cooled, and then digested with 1 ml hydrofluoric acid. Samples were placed back on the block digester for 45-60 minutes. Tubes were removed and brought to a 50 ml volume with distilled water. Stoppered tubes were shaken and then analyzed by inductively coupled plasma atomic emission spectrometry (Iris Intrepid II XSP, model 14463001; Thermo Electron Corporation, Franklin, MA) equipped with an ASX-520 autosampler. Data are expressed as ppm (Table 2).

2.3 Extraction and determination of phenolic content in the tissues of *P. oceanica*, and phenolic and volatile content of above-ground tissues of *Z. marina*

Total phenolic content for both species, and total volatile content for *Z. marina*, were determined to ascertain whether tissue collected from impacted (heavy metal pollution for both species) and control sites differed. A different method is used for the definition of the phenolic and volatile compounds, because the measurements were realized in different labs. For *P. oceanica*, extraction of total phenolic compounds was carried out on 0.5 g dry wt freeze-dried foliage leaf or sheath tissue. Extraction followed Cuny et al. (1995) and consisted of infusing each sample at 40°C in 50 % (v/v) aqueous ethanol in darkness for 3 h. The extract was acidified with a few drops of 2N HCl, the ethanol was evaporated under vacuum, and the aqueous residue extracted with ethanol/acetic acid. The organic phase was dried using anhydrous Na_2SO_4. Concentration of total phenolic compounds was measured by colorimetry (Swain & Hillis, 1959) using Folin-Denis reagent (Folin and Dennis, 1915). Phloroglucinol (Frantzis, 1992) was used for elaboration of standard curves. For *Z. marina*, phenolics were extracted using $MeOH/CH_2CH_2$ (50/50) from 200 mg dry wt of freeze dried above-ground tissue, filtered using VWR grade 415 filter paper, and blown dry using nitrogen gas to prevent oxidation. After resdisolving in $MeOH/CH_2CH_2$ (50/50), the extract was again filtered, placed in an auto-sampler vial (Chromatography Research Supplies, Addison, IL) and injected into a high pressure liquid chromatograph (HPLC) (HP Model 1100; Agilent 1100 Series, Model G1313A; Santa Clara, CA) equipped with a diode-array detector (Model G1316A) and a C_{18} reverse phase 5μm column (Phenomenex, Torrance, CA). The HPLC solvents were A = water/acetic acid (98:2); B = acetonitrile/acetic acid (98:2). Temperature was 50°C, flow rate 1ml/min, and wavelength of the detector set at 280 nm (optimized for *Z. marina* phenolic compounds). Phenolic content is expressed as total peak height /200 mg dry wt. To obtain volatile compound content in *Z. marina* samples, 3 g fresh wt of above-ground tissue was ground to a fine powder in liquid nitrogen and hexane. The extract was then filtered, and the filtrate injected into a capillary gas chromatograph (HP Model 6890) equipped with a head-space sampler (Perkin-Elmer HS 40 XL; Waltham, MA) and a HP-1 column. Oven temperature was 80°C, needle temperature 85°C, transfer temperature 120°C, thermostat time 10 min, pressurizing time 0.6 min, injection time 0.2

min, and withdrawal time 0.5 min. The ramp GC program was 40-210ᵒC at intervals of 3ᵒC ramp/min. Total volatile compound content is expressed as total peak height per 3 g fresh wt tissue.

2.4 Statistical analysis

Data from *P. oceanica* samples were analyzed using a three-way ANOVA to allow comparisons between the phenolic compounds and mercury levels according to tissue, site and sampling period. Since the interaction among these factors was significant, one-way analyses followed by a Tukey test (for analyses over the annual cycle) or Student-t test (for analyses of tissue and site factors at given months) were performed (Zar, 1999). Normality and homoscedasticity were verified by Shapiro Wilks and Bartlett tests, respectively (Zar, 1999). The relationships between phenolic compounds and mercury level were assessed using correlation and regression analyses in Statgraphics plus (ver 3.1) for Windows. Data from *Z. marina* are expressed as ppm for heavy metals, total peak area per 200 mg freeze dried tissue for phenolics, and total peak area per 3 g fresh wt for volatiles. Since all samples were randomly collected along a transect, each sample is treated as an independent experimental unit. Comparison of heavy metal content between impacted and control sites in *Z. marina* above- and below-ground tissues, and for phenolic and volatile content in above-ground tissues, was conducted using a one-way ANOVA, SAS GLM program (SAS, 1996).

3. Results

3.1 Site and tissue differences in heavy metal contamination

Foliage leaf and sheath tissue of *P. oceanica* from the industrially impacted Rosignano site showed large and significant ($p < 0.05$) differences in mercury content when compared to the control Tonnara site (Table 1).

Tissue Type		Mercury impacted site (Rosignano)	Control site (Tonnara)
Foliage Leaves	June 2000	233 ± 23	77 ± 11
	January 2001	317 ± 41	79 ± 15
Sheaths	June 2000	368 ± 26	64 ± 8
	January 2001	215 ± 16	80 ± 19

Table 1. Mercury levels (ng/g dw) in foliage leaf and sheath tissues of *P. oceanica* collected at different sites and different sampling periods.

Samples of above-ground tissue collected in April 2000 from *Z. marina* plants growing in the impacted site were higher in iron, aluminium, and copper when compared to tissue from the control site (Table 2). However, above-ground tissue from the control site was significantly higher in zinc, nickel, molybdenum, and mercury when compared to the impacted site (Table 2). For the July 2000 samples, the only significant differences were that nickel and copper were in highest concentration in plants from the impacted site when compared to plants from control site (Table 2). For below-ground tissue of *Z. marina* in April, samples from the industrially impacted site were significantly higher ($p < 0.05$) for iron, aluminium, nickel, manganese, copper, cadmium, chromium, and lead when compared to the control site (Table 2). None of the heavy metals was higher in concentration in the control site for samples taken in April 2000. For the July 2000 samples, barium, iron, aluminium, zinc, manganese, copper, cadmium, arsenic, and chromium were higher in the plants from the impacted site when compared to the control site, and cobalt and strontium were higher in plants from the control site (Table 2).

Heavy Metals	Site (ppm)*							
	Above-ground Tissue				Below-ground Tissue			
	April		July		April		July	
	Industrially impacted site	Control site	Industrially impacted site	Control site	Industrially impacted site	Control site	Industrially impacted site	Control site
Barium	323(41)ᵃ	364(32)ᵃ	279(107)ᵃ	312(55)ᵃ	466(136)ᵃ	420(100)ᵃ	570(148)ᵃ	315(84)ᵇ
Iron	320(127)ᵃ	180(58)ᵇ	204(87)ᵃ	142(94)ᵃ	5801(2846)ᵃ	1068(540)ᵇ	5591(1503)ᵃ	576(263)ᵇ
Aluminum	183(75)ᵃ	119(43)ᵇ	88(42)ᵃ	100(81)ᵃ	1626(1341)ᵃ	665(435)ᵇ	1737(494)ᵃ	503(336)ᵇ
Zinc	100(11)ᵃ	119(13)ᵇ	102(22)ᵃ	110(15)ᵃ	134(45)ᵃ	133(44)ᵃ	169(46)ᵃ	96(16)ᵇ
Nickel	55(21)ᵃ	104(25)ᵇ	45(16)ᵃ	23(15)ᵇ	127(78)ᵃ	34(13)ᵇ	63(20)ᵃ	64(31)ᵃ
Manganese	37(6)ᵃ	42(6)ᵃ	48(16)ᵃ	51(7)ᵃ	38(33)ᵃ	11(6)ᵇ	26(7)ᵃ	10(4)ᵇ
Copper	14(2)ᵃ	12(2)ᵇ	16(4)ᵃ	10(1)ᵇ	40(21)ᵃ	19(29)ᵇ	43(12)ᵃ	10(3)ᵇ
Molybdenum	5(2)ᵃ	6(1)ᵇ	7(2)ᵃ	8(1)ᵃ	0(0)	----**	0(0)ᵃ	0(1)ᵃ
Cadmium	2(1)ᵃ	1(1)ᵃ	2(1)ᵃ	2(1)ᵃ	13(8)ᵃ	4(2)ᵇ	11(3)ᵃ	3(1)ᵇ
Arsenic	4(2)ᵃ	3(2)ᵃ	3(2)ᵃ	3(2)ᵃ	8(7)ᵃ	8(1)ᵃ	10(6)ᵃ	1(2)ᵇ
Cobalt	2(1)ᵃ	4(1)ᵃ	2(1)ᵃ	2(1)ᵃ	2(1)ᵃ	1(1)ᵃ	1(1)ᵃ	2(1)ᵇ
Mercury	1(1)ᵃ	2(1)ᵇ	0(1)ᵃ	1(1)ᵃ	3(3)ᵃ	2(1)ᵃ	4(1)ᵃ	4(2)ᵃ
Strontium	1(2)ᵃ	2(3)ᵃ	4(3)ᵃ	4(2)ᵃ	3(4)ᵃ	6(4)ᵃ	0(0)ᵃ	1(2)ᵇ
Chromium	1(1)ᵃ	2(1)ᵃ	1(0)ᵃ	1(0)ᵃ	7(4)ᵃ	2(1)ᵇ	6(2)ᵃ	1(1)ᵇ
Lead	0(0)ᵃ	1(2)ᵃ	0(0)ᵃ	0(0)ᵃ	13(23)ᵃ	6(3)ᵇ	0(1)ᵃ	0(1)ᵃ

Table 2. Differences in accumulation of heavy metals in above- and below- ground tissues of *Z. marina* between impacted and control sites [April, July 2000; x, -]. *Means followed by different letters are significantly different at $p < 0.05$; Means followed by the same letter (i.e. "a") are not significantly different at $p < 0.05$. **Insufficient sample for analysis.

3.2 Production of phenolic and volatile compound content in plant tissues between impacted and control sites

Foliage leaves from Tonnara (20.5 mg.g^{-1}) were significantly higher (Tukey test, $p < 0.05$) in phenolic content in January 2001 compared to plants from the mercury impacted Rosignano site (13.2 mg.g^{-1}), but were not significantly different in the June 2000 samples (Fig. 1). For sheaths, the levels of total phenolic compounds from Tonnara plants in June and January (9.2 and 15.2 mg.g^{-1}, respectively) were significantly higher than those measured in plants at the Rosignano site (5.0 and 6.4 mg.g⁻¹, respectively) (Tukey test, $p<0.05$). Phenolic content was higher across sites and sampling times in *P. oceanica* foliage leaves compared to sheaths in all comparisons (Mann and Whitney test, $p> 0.05$).

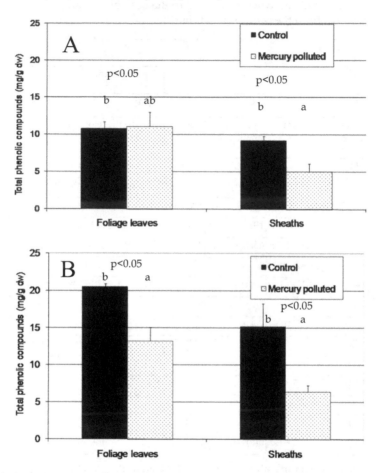

Fig. 1. Total phenolic concentration (mg.g⁻¹ dw) in foliage leaf and sheath tissues of *P. oceanica* in Tonnara (control) and Rosignano (mercury polluted) in June 2000 (A) and January 2001 (B).

For *Z. marina* total phenolic content in above-ground tissues collected from plants at the control site always was higher when compared to above-ground tissues collected from the impacted site for both April and July 2000 (Fig. 2). However, the only significant difference was in July where the control site produced a higher amount of total phenolic (65.8 vs 50.8 peak area / 200 mg dry wt, respectively; $p<0.05$). Total volatile compound production also was higher at both sampling periods, but the only significant difference occurred in the April 2000 sampling where above-ground tissues from the control site showed an average peak area of 551 per 200 mg dry wt tissue compared to 352 at the impacted site ($p<0.05$).

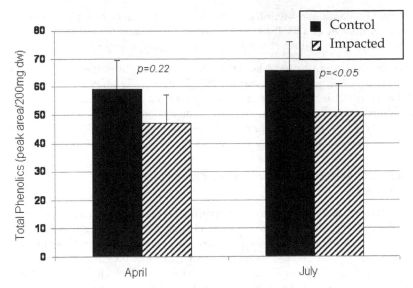

Fig. 2. Total phenolic content in above-ground tissue from Z. marina plants growing in heavy metal impacted and control sites (April and July, 2000).

4. Discussion

4.1 Tissue and site differences in heavy metal content

Results presented indicate that plant tissues of P. oceanica and Z. marina significantly accumulated high levels of heavy metals when growing on heavy metal-impacted sites (Tables 1 & 2). At the Rosignano site, when compared to the control Tonnara site, foliage leaves and sheaths contained two to over six times the amount of mercury. These patterns of accumulation are consistent with findings by other authors who have studied the same sites (Capiomont et al., 2000; Ferrat, 2001; Ferrat et al., 2003b; Maserti & Ferrara, 1991).

Z. marina plants from the heavy-metal impacted site accumulated significantly higher concentrations of iron, aluminum, nickel, and copper in their above-ground tissues when compared to the control site (Table 2). In addition, below-ground tissue of Z. marina plants from the industrially-impacted site accumulated over three, and up to five, times the levels of heavy metals compared to plants from the control site. A striking difference between above- and below-ground tissue, is that below-ground tissue from the impacted site accumulated 12 heavy metals (barium, iron, aluminum, zinc, nickel, manganese, copper, cadmium, arsenic, cobalt, chromium, lead; Table 2) while above-ground tissue only accumulated four heavy metals (iron, aluminum, nickel, copper) (Table 2). Another major difference is that the quantity of heavy metals accumulated in the below-ground tissue was higher for most of the heavy metals compared to that in the above-ground tissue.

Variation in metallic accumulation between above- and below-ground seagrass tissue has been discussed by various authors (see synthesis in Pergent Martini & Pergent, 2000), and could be a function of differences in binding sites or seasonal translocation between above- and below-ground structures (Libes & Boudouresque, 1987; Ward, 1987). The level of environmental contamination within a particular site also may be an important factor. For example Capiomont et al. (2000) found that mercury content was higher in the interstitial water than in the water column at our Rosignano sampling location.

Heavy metals are known to have adverse affects on the physiology of *P. oceanica* and *Z. marina* as well as other seagrasses (Ward, 1987). Lyngby & Brix (1984) have shown that the order of heavy metal inhibition of growth of *Z. marina* from greatest to least is mercury, copper, cadmium, zinc, chromium, and lead. Interestingly mercury was not significantly accumulated by *Z. marina* at our impacted site but the other five generally followed the pattern described by Lynby & Brix (1984) (Table 2).

4.2 Phenolic and volatile compound production in plant tissues between impacted and control sites

Our results suggest that total phenolic compound levels within seagrass tissue could be an indicator of site quality. Differences in production of phenolics in tissues from both species were noted between impacted and control sites. For foliage leaves and sheaths of *P. oceanica* collected in January, and above-ground tissue of *Z. marina* collected in July, total phenolic content was significantly lower in plants collected from industrial sites (Fig. 1 & 2). This is supported by Vergeer et al. (1995) who concluded that a decrease of total phenolic compounds in the tissue of *Z. marina* indicated plants may be growing in unsuitable environmental conditions. Noteworthy is that correlation analysis indicated a significant ($p < 0.05$) inverse relationship between heavy metal content and the health of plants as measured by phenolic content for *P. oceanica* ($r^2 = 69.8$ %, linear model of regression: mercury $= 0.22 - 0.0055 *$ phenol for sheaths).

Additionally, gas chromatography analysis of volatile compounds from *Z. marina* indicated that above-ground tissue from plants growing in the impacted site was significantly lower in volatiles from the April collection, when growth begins in Northern Puget Sound (Phillips, 1984) compared to tissue from the control site (Fig. 3). However, no significant differences occurred in volatile compound production between impacted and control sites in the July collection.

4.3 Phenolic compound production with regard to tissue and time collection

For *P. oceanica*, the concentration of phenolic compounds differed between foliage leaves and sheaths being higher in leaf tissue regardless of site. Similarly, Agostini et al. (1998) found higher concentrations (6 mg.g^{-1}) in the apical parts and youngest leaves and lower concentrations in sheaths (0.1 mg.g^{-1}). Also, in our study significant variation was observed between seasons; for example, phenolic levels were found to be higher in the January 2001 samples compared to the June 2000 samples.

Differences occur in the natural products analyzed depending on month of collection for both *P. oceanica* and *Z. marina* *(Fig. 1-3)*. For example, *P. oceanica* foliage leaves and sheaths in January 2001 were higher in phenolic content than those collected in June 2000 (Fig. 1). While phenolic content in above-ground *Z. marina* tissue was similar in concentration between April and July (Fig. 2), but volatile compounds in above-ground tissue collected in April were significantly higher than those collected in July (Fig. 3). April and July were selected as sampling times for *Z. marina* because they represent early and mature tissue growth in the Northern Puget Sound (Phillips, 1984). However, in a preliminary study in which *Z. marina* shoots were collected in February 2000, plants from the heavy metal-impacted site produced only 19% of the total phenolic content when compared to plants from the control site (Zou et al., unpublished data). In order to establish when phenolics and volatiles may best indicate plant health, experimental designs need to involve sampling plants every two months throughout the year.

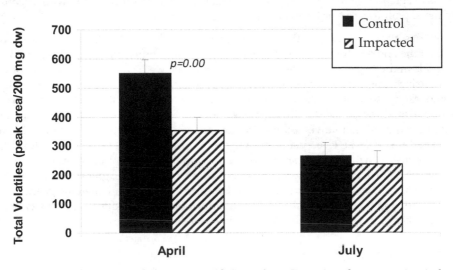

Fig. 3. Total volatile content of above-ground tissue from *Z. marina* plants growing in heavy metal impacted and control sites (April and July, 2000).

Finally, based on the response of different seagrass genotypes to disturbance (e.g. Ehlers et al., 2008; Hughes & Stachowicz, 2009; Wyllie-Echeverria et al., 2010), we suspect that variation in the type and concentration of heavy metal uptake may exist within different genotypes. However, this aspect of heavy metal accumulation in needs investigation in controlled conditions with seagrass species from different locations.

5. Conclusions

Significant differences were found in the accumulation of mercury in leaf and sheath tissues of *P. oceanica* when plants were growing on impacted sites as compared to sites not impacted heavily by mercury (Table 1). *Z. marina* plants growing in a site impacted by heavy metals associated with industrial pollution accumulated significantly higher amounts of iron, aluminium, nickel, and copper in above-ground tissues as compared to a non-impacted site, and higher amounts of barium, iron, aluminium, zinc, nickel, manganese, copper, cadmium, arsenic, chromium, and lead in below-ground tissues at the impacted site (Table 2). For *P. oceanica,* total phenolics were significantly higher in leaves at the control site when compared to the mercury impacted site for the January sampling period (Fig. 1). For sheath tissue total phenolics from the control site were significantly higher when compared to the mercury impacted site for both sampling periods (Fig. 1). For *Z. marina,* total phenolic content was higher in both sampling periods at the non-impacted site compared to the control site, but only significantly so for the July 2000 sampling period (Fig. 2). Total volatile content also was higher at the control site for both sampling periods, but only significantly higher for the April sampling period (Fig. 3). These results support the hypotheses that *P. oceanica* and *Z. marina* accumulate significant amounts of heavy metals from impacted sites, and that these accumulations are associated with reduced total phenolic and volatile compound content. Based on these supportive data, we conclude that *P. oceanica* and *Z. marina* are potential candidates as bio-surveillance organisms especially with regard to heavy metal pollution of coastal and estuarine ecosystems.

Since we observed variation in the production of phenolics and volatiles with regard to sampling time and season, a priority is the identification of individual phenolic and volatile compounds in the tissue of these two species. In our labs we have identified in one or both species using gas chromatography/mass spectroscopy and high pressure liquid chromatography several cinnamic acid and benzoic acid derivatives; these results are comparable to those found by Quackenbush et al. (1986). Additionally, these analyses indicate not only a quantitative decrease in total phenolic and volatile compounds, but also qualitative differences between plants growing on impacted and non-impacted sites (Ferrat et al., unpublished data for *P. oceanica*; Zou et al., unpublished data for *Z. marina*). Finally, since various environmental perturbations may adversely affect seagrass health (impact of human activity reviewed in Short & Wyllie-Echeverria, 1996), and thereby phenolic and volatile compound production, collaboration among scientists working at a diversity of sites would greatly facilitate progress toward this bio-surveillance effort.

6. Acknowledgments

We thank Tina, Victoria, Rebecca, and Tessa Wyllie-Echeverria and Carl Young for assistance with sample collection in Washington State. RGC and JZ thank the Department of Botany for the professional development funds supporting this research, and RGC is grateful for additional funding from the Karl G. Maeser Research Award. SWE gratefully acknowledges support from the Russell Family Foundation during the writing portion of this investigation.

7. References

Agostini, S.; Desjobert, J.M. & Pergent, G. (1998). Distribution of phenolic compounds in the seagrass *Posidonia oceanica*. *Phytochemistry*, Vol.48, No.4, (June 1998), pp. 611-617, ISSN 0031-9422

Augier, H. (1985). L'herbier à *Posidonia oceanica*, son importance pour le littoral Méditerranéen, sa valeur comme indicateur biologique de l'état de santé de la mer, son utilisation dans la surveillance du milieu, les bilans écologiques et les études d'impact. *Vie Marine*, Vol.7, (1985), pp. 85-113

Blandin, P. (1986). Bioindicateurs et diagnostic des systèmes écologiques. *Bulletin d'Ecologie*, Vol.17, No.4, (April 1986), pp. 211-307, ISSN 0395-7217

Brackup, I. & Capone, D.G. (1985). The effect of several metal and organic pollutants on nitrogen-fixation (acetylene reduction) by the roots and rhizomes of *Zostera marina* L. *Environmental and Experimental Botany*, Vol.25, No.2, (May 1985), pp. 145-151, ISSN 0098-8472

Brix, H.; Lyngby, J.E. & Schierup, H.H. (1983). Eelgrass (*Zostera marina* L.) as an indicator organism of trace metals in the Limfjord, Denmark. *Marine Environmental Research*, Vol.8, No.3, (March 1983), pp. 165-181, ISSN 0141-1136

Brix, H. & Lyngby, J.E. (1984). A survey of the metallic composition of *Zostera marina* L. in the Limfjord, Denmark. *Archiv für Hydrobiologie*, Vol.99, No.3, (March 1984), pp. 347-359, ISSN 0003-9136

Brix, H. (1997). Do macrophytes play a role in constructed treatment wetlands? *Water Science and Technology*, Vol.35, No.5, (May 1997), pp. 11-17, ISSN 0273-1223

Bryant, J.P.; Chapin F.S. & Klein, D.R. (1983). Carbon/nutrient balance of boreal plants in relation to vertebrate herbivory. *Oikos*, Vol.40, No.3, (March 1983), pp. 357-368, ISSN 0030-1299

Capiomont, A.; Piazzi, L. & Pergent, G. (2000). Seasonal variations of total mercury in foliar tissues of *Posidonia oceanica*. *Journal of the Marine Biological Association of the United Kingdom*, Vol.80, No.6, (June 2000), pp. 1119-1123, ISSN 0025-3154

Cates, R.G. (1996). The role of mixtures and variation in the production of terpenoids in conifer-insect-pathogen interactions, In: *Phytochemical Diversity and Redundancy in Ecological Interactions, Serie Recent Advances in Phytochemistry*, Romeo, J.T.; Saunders, J.A.; Barbosa, P., Vol.30, pp. 179-216, Plenum Press, ISBN 978-0-306-45500-1, New York

Chen, C.; Arjomandi, M.; Balmes, J.; Tager, I. & Holland, N. (2007). Effects of chronic and acute ozone exposure on lipid peroxidation and antioxidant capacity in healthy young adults. *Environmental Health Perspectives*, Vol.115, No.12, (December 2007), pp. 1732-1737, ISSN 0091-6765

Croteau, R. & Gershenzon, J. (1994). Genetic control of monoterpene biosynthesis in mints (*Mentha*: Lamiaceae), In: *Genetic Engineering of Plant Secondary Metabolism, Serie Recent Advances in Phytochemistry*, Ellis, B.E.; Kuroki, G.W.; Stafford, H.A., Vol.28, pp. 193-229, Plenum Press, ISBN 978-0-306-44804-1, New York

Croteau, R. & Sood, V. (1985). Metabolism of monoterpenes. Evidence for the function of monoterpenes and catabolism in peppermint (*Mentha piperita*). *Plant Physiology*, Vol.77, No.4, (April 1985), pp. 801-806, ISSN 0032-0889

Cuny, P.; Serve, L.; Jupin, H. & Boudouresque, C.F. (1995). Water soluble phenolic compounds of the marine phanerogam *Posidonia oceanica* in a Mediterranean area colonized by the introduced chlorophyte *Caulerpa taxifolia*. *Aquatic Botany*, Vol.52, No.3, (December 1995), pp. 237-242, ISSN 0304-3770

den Hartog, C.D. (1970). *The Seagrasses of the World*, Verhand Koninklijke Nederl Akad. Wetenschap Afd. Nat. Tweede Reeks, North-Holland Publication, Amsterdam

Dennison, W.C.; Orth, R.J.; Moore, K.A.; Stevenson, J.C.; Carter, V.; Kollar, S.; Bergstrom, P.W. & Batiuk, R.A. (1993). Assessing water quality with submerged aquatic vegetation: Habitat requirements as barometers of Chesapeake Bay health. *BioScience*, Vol.43, No.2, (February 1993) pp. 86-94, ISSN 0006-3568

Ehlers, A.; Worm, B. & Reusch, T.B.H. (2008). Importance of genetic diversity in eelgrass *Zostera marina* for its resilience to global warming. *Marine Ecology Progress Series*, Vol.355, (February 2008), pp. 1-7, ISSN 0171-8630

Emmons, C.L.; Peterson, D.M. & Paul, G.L. (1999). Antioxidant capacity of oat (Avena sativa L.) extracts. 2. In vitro antioxidant activity and contents of phenolic and tocol antioxidants. Journal of Agricultural and Food Chemistry, Vol.47, (December 1999), pp. 4894-4898, ISSN 0021-8561

Fierer, N.; Schimel, J.P.; Cates, R.G. & Zou, J. (2001). Influence of balsam poplar tannin fractions on carbon and nitrogen dynamics in Alaskan taiga floodplain soils. *Soil Biology and Biochemistry*, Vol.33, No.12-13, (October 2001), pp. 1827-1839, ISSN 0038-0717

Ferrat, L. (2001). *Réactions de la phanérogame marine Posidonia oceanica en réponse à des stress environnementaux*, Thèse de doctorat, Université de Corse, France

Ferrat, L.; Pergent-Martini C. & Roméo M. (2003a). Assessment of the use of biomarkers in aquatic plants for the evaluation of environmental quality: application to seagrasses. *Aquatic Toxicology*, Vol.65, No.2, (October 2003), pp. 187-204, ISSN 0166-445X

Ferrat, L.; Gnassia-Barelli, M.; Pergent-Martini, C. & Roméo, M. (2003b). Mercury and non-protein thiol compounds in the seagrass *Posidonia oceanica*. *Comparative Biochemistry and Physiology c*, Vol.134, No.1, (January 2003), pp. 147-155, ISSN 0742-8413

Ferrara, R.; Maserti, B.E. & Paterno, P. (1989). Mercury distribution in marine sediment and its correlation with the *Posidonia oceanica* prairie in a coastal area affected by a chlor-alkali complex. *Toxicological and Environmental Chemistry*, Vol.22, No.1-4, (April 1989), pp. 131-134, ISSN 0277-2248

Folin, O. & Denis, W. (1915). A colorimetric method for the determination of phenols (and phenol derivatives) in urine. *Journal of Biological Chemistry*, Vol.22, (1915), pp. 305-308, ISSN 0021-9258

Frantzis, A. (1992). *Etude expérimentale des niveaux de consommation et d'utilisation des macrophytes et des détritus dérivés par deux invertébrés benthiques Paracentrotus lividus et Abra ovata*, Thèse de doctorat, Université Aix Marseille II, France

Gershenzon, J. (1984). Changes in the level of plant secondary metabolite production under water and nutrient stress. In: *Phytochemical adaptation to stress, Serie Recent Advances in Phytochemistry*, Loewus F.A.; Timmermann B.N.; Steelink C., Vol.24, pp. 273-320, Plenum Press, ISBN 0306417200, New York

Giraud, G. (1977). *Contribution à la description et à la phénologie quantitative des herbiers à Posidonia oceanica (L.) Delile*, Thèse de Doctorat 3ème cycle, Université Aix-Marseille II, France

Goldberg, E.D.; Koide, M. & Hodeg, V. (1983). U.S. Mussel watch: 1977-1978 results on trace metals and radionuclides. *Estuarine Coastal and Shelf Science*, Vol.16, No.1, (January 1983), pp. 69-93, ISSN 0272-7714

Goodwin, T.W. & Mercer, E.I. (1983). *Introduction to Plant Biochemistry*, Pergamon Press, Oxford, ISBN 0080249221 9780080249223 0080249213 9780080249216, England

Green, E.P. & Short, F.T. (2003). *World Atlas of Seagrasses*, UNEP World Conservation Monitoring Centre, University of California Press, Berkeley, ISBN 0-520-24047-2, USA

Hadacek, F. (2002). Secondary Metabolites as Plant Traits: Current Assessment and Future Perspectives. *Critical Reviews in Plant Sciences*, Vol.21, No.4, (April 2002), pp. 273-322, ISSN 0735-2689

Hartman, T. (2007). *From waste products to ecochemicals: fifty years research of plant secondary metabolism*. Phytochemistry, Vol.68, No.22-24, (November-December 2007), pp. 2831-2846, ISSN 0031-9422

Heglmeier, A. & Zidorn, C. (2010). Secondary metabolites of *Posidonia oceanica* (Posidoniaceae). *Biochemical Systematics and Ecology*, Vol.38, No.5, (May 2010), pp. 1-7, ISSN 03051978

Hemminga, M. & Duarte, C. (2000). *Seagrass Ecology*, Cambridge University Press, ISBN 0-521-66184-6, USA

Horner, J.D.; Gosz, J.R. & Cates, R.G. (1988). *The role of carbon-based plant secondary metabolites in decomposition in terrestrial ecosystems.* American Naturalist, Vol.132, No.6, (June 1988), pp. 869-883, ISSN 0003-0147

Hughes, A.R. & Stachowicz, J.J. (2009). Ecological impacts of genetic diversity in the clonal seagrass *Zostera marina. Ecology*, Vol.90, No.5, (May 2009), pp. 1412-1419, ISSN 0012-9658

Kuo, J. & den Hartog, C. (2006). Seagrass morphology, anatomy and ultrastructure, In: *Seagrasses: Biology, Ecology, and Conservation,* Larkum, A.W.D.; Orth, R.J.; Duarte, C.M., pp. 51-87, Springer-Verlag, ISBN 978-1402029424, The Netherlands

Lafabrie, C.; Pergent, G.; Kantin, R.; Pergent-Martini, C. & Gonzalez, J-L. (2007). Trace metals assessment in water, sediment, mussel and seagrass species – Validation of the use of *Posidonia oceanica* as a metal biomonitor. *Chemosphere*, Vol.68, No.11, (August 2007), pp. 2033-2039, ISSN 0045-6535

Lange, C.R. & Lambert, K.E. (1994). Biomonitoring. *Water Environmental Research*, Vol.66, No.4, (June 1994), pp. 642-651, ISSN 1061-4303

Langenheim, J. (1994). Higher plant terpenoids: A phytocentric overview of their ecological roles. *Journal of Chemical Ecology*, Vol.20, No.6, (June 1994), pp. 1223-1280, ISSN 0098-0331

Libes, M. & Boudouresque C.F. (1987). Uptake and long-distance transport of carbon in the marine phanerogram *Posidonia oceanica. Marine Ecology Progress Series*, Vol.38, (June 1987), pp. 177–186, ISSN 0171-8630

Lipkin, Y.; Beer S. & Zakai D. (2003). The seagrases of the Eastern Mediterranean and the Red Sea, In: *World Atlas of Seagrasses,* Green, E.P.; Short F.T., pp. 65-73, UNEP World Conservation Monitoring Centre, University of California Press, Berkeley, ISBN 0-520-24047-2, USA

Lyngby, J.E. & Brix, H. (1984). The uptake of heavy metals in eelgrass *Zostera marina* and their effect on growth. *Ecological Bulletins*, Vol.36, (1984), pp. 81-84

Macheix, J.J. (1996). Les composés phénoliques des végétaux : quelles perspectives à la fin du XXème siècle? *Acta Botanica Gallica*, Vol.143, No.6, (June 1996), pp. 473-479, ISSN 1253-8078

Maserti, B.E. & Ferrara, R. (1991). Mercury in plants, soil and atmosphere near a chlor-alkali complex. *Water, Air, Soil Pollution*, Vol.56, No.1, (April 1991), pp. 15-20, ISSN 1567-7230

Maserti, B.E.; Ferrara, R. & Morelli, M. (1991). *Posidonia oceanica*: uptake and mobilization of mercury in the Mediterranean basin. In: *Proceedings of the FAO/UNEP/IAEA Worshop on the biological effects of pollutants on marine organisms*, Gabrielides, G.P., Vol. 59, pp. 243–249, MAP Technical Reports Series, Athens, Greece

McCarthy, J.F. & Shugart, L. (1990). *Biomarkers of Environmental Contamination*, Lewis Publishers: Boca Raton, ISBN 0873712846, Florida, USA

Molinier, R. & Picard, J. (1952). Recherches sur les herbiers de phanérogames marines du littoral méditerranéen français. *Annales de l'Institut Océanographique*, Vol.27, No.3, (1952), pp. 157-234, ISSN 0078-9682

Orth, R.J.; Carruthers, T.J.B.; Dennision, W.C.; Duarte, C.M.; Fourqurean, J.W.; Heck, Jr.K.L.; Hughes, A.R.; Kendrick, G.A.; Kenworthy, W.J.; Olyarnik, S.; Short, F.T.; Waycott, M. & Williams, S.L. (2006). A global crisis for seagrass ecosystems. *Bioscience*, Vol.56, No.12, (December 2006), pp. 987-996, ISSN 0006-3568

Osborn, D. & Datta A. (2006). Institutional and policy cocktails for protecting coastal and marine environments from land-based sources of pollution. *Ocean and Coastal Management*, Vol.49, No.9-10, (Januray 2006), pp. 576-596, ISSN 0964-5691

Pergent, G. (1991). Les indicateurs écologiques de la qualité du milieu marin en Méditerranée. *Oceanis*, Vol.17, No.4, (April 1991), pp. 341-350, ISSN 0182-0745

Pergent-Martini, C. (1998). *Posidonia oceanica*: a biological indicator of past and present mercury contamination in the Mediterranean Sea. *Marine Environmental Research*, Vol.45, No.2, (March 1998), pp. 101-111, ISSN 0141-1136

Pergent-Martini, C. & Pergent G. (2000). Are marine phanerogams a valuable tool in the evaluation of marine trace-metal contamination: example of the Mediterranean sea? *International Journal of Environment and Pollution*, Vol.13, No.1-6, (January 2000), pp. 126-147, ISSN 0957-4352

Pergent-Martini, C.; Pergent, G.; Fernandez, C. & Ferrat, L. (1999). Value and use of *Posidonia oceanica* as a biological indicator, In: *Proceedings MEDCOAST-EMECS 99 Joint Conference Land-ocean interactions: managing coastal ecosystems*, Vol.1, pp. 73-90, MEDCOAST, Middle East Technical University Publication, ISBN 975-429-142-X, Ankara, Greece

Phillips, D. (1992). Flavonoids: plant signals to soil microbes. In: *Phenolic Metabolism in Plants, Serie Recent Advances in Phytochemistry*, Stafford, H.; Ibrahim, R., Vol.26, pp. 201-231, Plenum Press, ISBN 9780306442315, New York, USA

Phillips, R.C. (1984). *The ecology of eelgrass meadows in the Pacific Northwest: A community profile*, FWS/OBS-84/24, U.S. Dept. of the Interior, Washington, USA

Prange, J.A. & Dennison, W.C. (2000). Physiological responses of five seagrass species to trace metals. *Marine Pollution Bulletin*, Vol.41, No.7-12, (July 2000), pp. 327-336, ISSN 0025-326X

Procaccini, G.; Buia, M.C.; Gambi, M.C.; Perez, M.; Pergent, G.; Pergent-Martini, C. & Romero, J. (2003). The seagrasses of the Western Mediterranean, In: *World Atlas of Seagrasses*, Green, E.P.; Short F.T., pp. 48-58, UNEP World Conservation Monitoring Centre, University of California Press, Berkeley, ISBN 0-520-24047-2, USA

Quackenbush, R.C.; Bunn, D. & Lingren, W. (1986). HPLC determination of phenolic acids in the water-soluble extract of *Zostera marina* L. (eelgrass). *Aquatic Botany*, Vol. 24, No.1, (January 1986), pp. 83-89, ISSN 0304-3770

Ragan, M.A. & Glombitza, K.W. (1986). Phlorotannins, brown algal polyphenols. In: *Progress in Phycological Research*, Round, F.E.; Chapman D.J., Vol.4, pp. 130-241, Elsevier Biomedical Press, ISBN 094873700X, 9780948737008, UK

Rainbow, P.S. & Phillips, D.J.H. (1993). Cosmopolitan biomonitors of trace metals. *Marine Pollution Bulletin*, Vol.26, No.11, (November 1993), pp. 593-601, ISSN 0025-326X

Ramade, F. (1992). *Précis d'écotoxicologie*, Masson, Collection d'écologie, ISBN 2-225-82578-5, France

Reichert, W.L.; Myers, M.S.; Peck-Miller, K.; French, B.; Anulacion, B. F.; Collier, T. K.; Stein, J.E. & Varanasi, U. (1998). Molecular epizootiology of genotiocic events in marine fish: Linking contaminant exposure, DNA damage, and tissue-level alterations. *Mutation Research*, Vol.411, No.3, (November 1998) pp. 215-225, ISSN 0027-5107

Rice–Evans, C.A.; Miller, N.J.; Bolwell, P.G.; Bramley, P.M. & Pridham, J.B. (1995). The relative antioxidant activities of plant-derived polyphenolic flavonoïds. *Free Radical Research*, Vol.22, No.4, (April 1995), pp. 375-383, ISSN 1071-5762

SAS Institute, Inc. (1996). *SAS User's Guide: Statistics*, Carey, ISBN 0-917382-01-3, USA

Schimel, J.P.; Cates, R.G.; Clausen, T.P. & Reichardt, P.B. (1996). Effects of balsam poplar (*Populus balsamifera*) tannins and low molecular weight phenolics on microbial activity in taiga floodplain soil: Implications for changes in N cycling during succession. *Canadian Journal of Botany*, Vol.74, No.1, (January 1996), pp. 84-90, ISSN 0008-4026

Short, F.T. & Wyllie-Echeverria, S. (1996). Human-induced and disturbance in seagrasses. *Environmental Conservation*, Vol.23, No.1, (January 1996), pp. 17-27, ISSN 0376-8929

Stephensen, E.; Savarsson, J.; Sturve, J.; Ericson, G.; Adolfsson-Erici, M. & Forlin, L. (2000). Biochemical indicators of pollution exposure in shorthorn sculpin (*Myoxocephalus scorpius*), caught in four harbours on the southwest coast of Iceland. *Aquatic Toxicology*, Vol.48, No.4, (April 2000), pp. 431-442, ISSN 0166-445X

Swain, T. & Hillis, W.E. (1959). *The phenolic constituents of* Prunus domestica. *I. The quantitative analysis of phenolic constituents.* Journal of the Science of Food and Agriculture, Vol.10, No.1, (1959), pp. 63-68, ISSN 0022-5142

Swain, T. (1977). Secondary compounds as protective agents. *Annals Review of Plant Physiology*, Vol.28, (1977), pp. 479-501, ISSN 0066-4294

Tett, P.; Gowen, R.; Mills, D.; Fernandes, T.; Gilpin, L.; Huxham, M.; Kennington, K.; Read, P.; Service, M.; Wilkinson, M. & Malcolm, S. (2007). Defining and detecting undesirable disturbance in the context of marine eutrophication. *Marine Pollution Bulletin*, Vol.55, No.1-6, (2007), pp. 282-297, ISSN 0025-326X

Tomlinson, P.B. (1974). Vegetative morphology and meristem dependence: The formation of productivity in seagrasses. *Aquaculture*, Vol.4, (1974), pp. 107-130, ISSN 0044-8486

Vergeer, L.H.T.; Aarts, T.L. & Groot, J.D. (1995). The wasting disease and the effect of abiotic factors (light intensity, temperature, salinity) and infection with *Labyrinthula zosterae* on the phenolic content of *Zostera marina* shoots. *Aquatic Botany*, Vol.52, No.1, (September 1995), pp. 35-44, ISSN 0304-3770

Vangronsveld, J.; Mocquot, B.; Mench, M. & Clijsters, H. (1997). Biomarqueurs du stress oxydant chez les végétaux, In: *Biomarqueurs en écotoxicologie - Aspects fondamentaux*, Lagadic L.; Caquet T.; Amiard J.C.; Ramade F., pp. 165-181, Dunod, ISBN 2225830533, Paris, France

Verges, A.; Becerro, M.A.; Alcoverro, T. & Romero, J. (2007). Experimental evidence of chemical deterrence against multiple herbivores in the seagrass *Posidonia oceanica*. *Marine Ecology Progress Series*, Vol.343, (August 2007), pp. 107-114, ISSN 0171-8630

Ward, T.J. (1987). Temporal variation of metals in the seagrass *Posidonia australis* and its potential as a sentinel accumulator near a lead smelter. *Marine Biology*, Vol.95, No.2, (February 1987), pp. 315-321, ISSN 0025-3162

White, C.S. (1986). Volatile and water-soluble inhibitors of nitrogen mineralization and nitrification in a ponderosa pine ecosystem. *Biology and Fertility of Soils*, Vol.2, (1986), pp. 97-104, ISSN 0178-2762

Wyllie-Echeverria, S. & Ackerman J.D. (2003). The seagrasses of the Pacific Coast of North America, In: *World Atlas of Seagrasses*, Green, E.P.; Short F.T., pp. 199-206, UNEP World Conservation Monitoring Centre, University of California Press, Berkeley, ISBN 0-520-24047-2, USA

Wyllie-Echeverria, S.; Talbot S.L. & Rearick J.R. (2010). Genetic structure and diversity of *Zostera marina* (eelgrass) in the San Juan Archipelago, Washington, USA. *Estuaries and Coasts*, Vol.33, No.4, (July 2010), pp. 811-827, ISSN 1559-2723

Zapata, O. & Mc Millan, C. (1979). Phenolic acids in seagrasses. *Aquatic Botany*, Vol.7, (1979), pp. 307-317, ISSN 0304-3770

Zar, J.H. (1999). *Biostatistical analysis*, Prentice-Hall International, ISBN 9780130815422, U.K

9

Interplay of Physical, Chemical and Biological Components in Estuarine Ecosystem with Special Reference to Sundarbans, India

Suman Manna[1], Kaberi Chaudhuri[1], Kakoli Sen Sarma[1], Pankaj Naskar[1],
Somenath Bhattacharyya[1] and Maitree Bhattacharyya[2]*
1Institute of Environmental Studies and Wetland Management,
2Department of Biochemistry, University of Calcutta,
Kolkata
India

1. Introduction

An estuary is a partly enclosed coastal body of water with one or more rivers or streams flowing into it, and with a free connection to the open sea. (Jara-Marin et al., 2009; Crossland et al. 2005). They are the transition zones or ecotones between riverine and marine habitats, which differ both in abiotic and biotic factors (McLusky & Elliott, 2004) but many of their important physical and biological attributes are not transitional, but unique. These highly dynamic and rapidly changing systems form a complex mixture of many different habitat types. These habitats do not exist in isolation, but rather have physical, chemical and biological links between them, for example, in their hydrology, in sediment transport, in the transfer of nutrients and in the way mobile species move between them both in seasonally and single tidal cycles. The most distinctive feature that contrasts estuaries from other biomes is the nature and the variability of the physicochemical forces that influence this ecosystem. In contrast the low diversity, the estuarine ecosystems achieve very high productivities through the continuous arrival of new nutrients supply. They are very productive biomes and support many important ecosystem functions like biogeochemical cycling and movement of nutrients, mitigation of floods, maintenance of biodiversity and biological production (Patrick et. al., 2005). The estuarine environment is characterized by a constant mixing of freshwater, saline seawater, and sediment, which is carried into the estuary from the sea and land. The mixture and fluctuation of salt and freshwater impose challenges to the animals and microbes. Along the gradient of conditions from the open sea into the sheltered estuary the salinity ranges from full strength seawater to freshwater, and sedimentary conditions also varies from fine sediment to coarse sediments. Other changes include nutrient input, pollutant and chemical concentration along with estuarine flows (McLusky & Elliott, 2004). The productivity and variety of estuarine habitats support a wonderful abundance and diversity of species.

*Corresponding Author

The ecosystem of any estuary is dependent on both the natural processes (like tide, current, bathymetry, nutrient influx etc) as well as anthropogenic activities (like agriculture, aquaculture etc in the adjoining land part and/or the number and frequency of mechanized boats, trawlers plying within the estuary etc). The entire process is extremely complicated where a balance is achieved through interaction between different components, not clearly well understood so far in many estuaries. The movement of water mass and consequent circulation pattern within an estuary is dependent and thereby should be considered as a response to astronomical tides, inflow of fresh water (i. e. head ward discharge of the fresh water), winds, density of saline water and consequent stratification of different water column etc. At the same time, the basin topography (bathymetry), air-water interaction, water sedimentation interface, mixing characteristics, frictional loss at the bottom, and the rotational effects of the earth, together with the above mentioned driving forces, constitute an extremely complicated balance that conserves mass, momentum, energy, and conservative solutes in the system.

Estuaries are highly dynamic systems with large seasonal and spatial gradients of biogeochemical compounds and processes. Linking land to the ocean, they are often greatly influenced by human activities, including enhanced organic matter and nutrient loadings. Among other parameters, the balance between organic matter and nutrient loading is critical in determining the balance between autotrophy and heterotrophy at the ecosystem level (Kemp et. al., 1997). Estuarine dynamics has been well studied in temperate system such as Chesapeake Bay (Boynton et. al., 1982), San Francisco Bay (Cloern, 1996) and the Baltic sea (Conley et. al., 2000). Tropical and subtropical estuaries received comparatively less study but are experiencing noticeable anthropogenic alterations (Eyre, 1997). Sundarbans is an example of tropical, nutrient rich and turbid estuary.

2. Unique features of Sundarbans estuary

Sundarbans is the largest deltaic tidal halophytic mangrove forest in the world, (Blasco, 1977) with an area of 10, 200 sq km area, spreading over India (4263 sq km of Reserve forest) and Bangladesh (5937 sq km of Reserve forest) (Fig 1). Sundarbans, world's largest delta (80, 000 sq. km.) formed from sediments deposited by three great rivers, the Ganges, Brahmaputra and Meghna, which converges on the Bengal Basin. The area experiences a subtropical monsoon climate with annual rainfall of about 1600-1800 mm and several cyclonic storms (Manna et. al., 2010).

Indian Sundarbans is known as Hoogly-Matla estuary (Hooghly is the Lower part of River Ganges), where apart from Hoogly and Matla, there are innumerable big & small rivers criss-crossing the Sundarbans namely Bidya, Saptamukhani, Raimangal, Muriganga, Thakuran, Gomor etc. Many rivers have become almost completely cut off from the main freshwater sources (Sanyal & Bal, 1986) as for example Bidya, Matla are devoid of fresh water connection due to siltation in the upstream region and are converted into tidal creeks.

Sundarbans is intersected by a complex network of tidal waterways, mudflats and small islands of salt-tolerant mangrove forests. The waterways in this tiger reserve are maintained largely by the diurnal tidal flow (Lahiri, 1973). Sundarbans is known for the eponymous Royal Bengal Tiger (*Panthera tigris tigris*), as well as numerous fauna including species of birds, spotted deer, crocodiles and snakes.

Interplay of Physical, Chemical and Biological Components in Estuarine Ecosystem with Special
Reference to Sundarbans, India

185

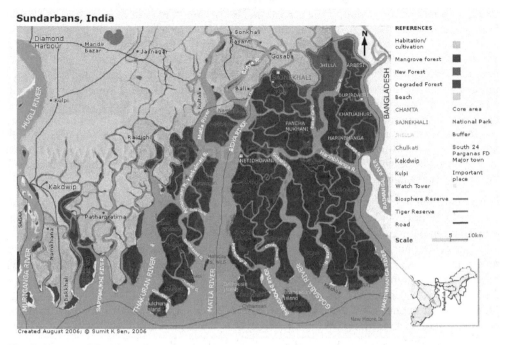

Fig. 1. Geographical location of Sundarban estuary, India

Sundarban delta has the distinction of encompassing the world's largest Mangrove Forest
belt and has been identified as World natural heritage site by UNESCO in 1974, National
park in 1984 and World heritage site by IUCN in 1989. The mangrove ecosystems in Indian
Sundarbans are well known not only for the aerial extent, but also for the species diversity
and richness. The biodiversity of Sundarbans includes about 350 species of vascular plants,
250 fishes and 300 birds, along with numerous species of phytoplankton, fungi, bacteria,
zooplankton, benthic invertebrates, molluscs, reptiles, amphibians and mammals (Gopal &
Chauhan, 2006) It is natural habitat of many rare and endangered species including the
Royal Bengal tiger ((*Panthera tigris*) and it is the only mangrove tiger land in the world.
Other important species are Estuarine Crocodile (*Crocodilus porosus*), Gangetic Dolphin
(*Platinista gangetica*), Snubfin dolphin (*Orcella brevirostris*), River Terrapin (Batagur baska),
Batagur baska, Pelochelys bibroni, Chelonia mydas., marine turtles like Olive Ridley (*Lepidochelys
olivacea*), Green Sea Turtle (*Chelonia mydas*), Hawksbill Turtle (*Eritmochelys imbricata*), thus
making it a natural biodiversity hot spot.

3. Mangrove of Sundarbans

Mangroves are a community of trees and shrubs growing in intertidal forested wetlands
restricted to the tropical and subtropical regions (Tomlinson, 1986). Total global area of
mangrove forest is estimated to only 18. 1 million ha (Spalding et. al., 1997), against over 570
million ha of freshwater wetlands including peat lands but excluding paddy fields (Spiers
et. al., 1999). Mangroves are the only woody halophytes dominated ecosystem situated at
the confluence of land and sea, they occupy a harsh environment, being daily subject to tidal

changes in temperature, water, salt exposure and varying degree of anoxia (Alongi, 2002). Mangrove forests are recognized as highly productive ecosystems that provide large quantities of organic matter to adjacent coastal waters in the form of detritus and live animals (Holguin et. al., 2001). They provide critical habitat for a diverse marine and terrestrial flora and fauna. Healthy mangrove forests are key to healthy marine ecology. They may be considered as self maintaining coastal, inter-tidal estuarine compartment, which thrives due to constant interaction with terrestrial and marine ecosystem. They are vital to coastal communities as they protect them from damage caused by tsunami waves, erosion, and storms and serve as a nursery for fish and other species that support coastal livelihoods. In addition, they have a staggering ability to sequester carbon from the atmosphere and serve as both a source and repository for nutrients and sediments for other inshore marine habitats.

Sundarbans mangrove estuarine ecosystem is one of the largest detritus-based ecosystems of the world (Pillay, 1958 and Ray, 2008). Litterfall of mangroves supplies the detritus and nutrients regulating the productivity of adjacent Hooghly–Brahmaputra estuarine complex which act as an important nursery ground for many commercially important shell and fin fishes. Due to large scale human intervention from the beginning of last century, several species have become extinct or are in very much threatened or degraded state (Gopal &Chauhan, 2006; Sodhi & Brok, 2007), The loss of the mangroves will have devastating economic and environmental consequence. Royal Bengal tiger, Javan rhino, wild buffalo, hog deer, and barking deer are on the verge of extinction. But any systematic approach towards studying the ecosystem dynamics of Sundarbans has not been attempted so far (Alongi, 2009;Gopal & Chauhan, 2006), where we have attempted to fill in the gaps and create a road map for the sustenance of this World heritage site.

4. Materials and methods

4.1 Physico-chemical analysis

4.1.1 Tidal velocity and current speed

Tide measurement was performed by Valeport MIDAS WTR non directional tide gauge serial no. 34890 (Valeport, U. K). The MIDAS WTR Wave Recorder uses the proven Linear Wave. The MIDAS WTR Wave Recorder uses the proven Linear Wave Theory wave analysis method of measurement. It has high accuracy piezo-resistive pressure sensors and a fast response PRT temperature sensor as standard. Current speed and direction was measured with Aanderra made Doppler Current Sensor 4420 Serial no. 282 Signal type CANbus.

4.1.2 Water quality parameters

Water temperature, pH and conductivity were measured in situ with Hach Portable Meters (HQ40d). Turbidity was measured by using portable turbidity meter (Hach 2100P), salinity was determined in practical salinity units by Knudsen method (Knudson, 1901), dissolved oxygen concentration was studied according to the method of Winkler (JGFOS Protocol, 1994) nutrients like inorganic nitrogen (ammonia, nitrite, nitrate, total nitrogen), soluble phosphate, total phosphate, and reactive silicate were measured according to the same methodology (JGFOS Protocol, 1994).

Interplay of Physical, Chemical and Biological Components in Estuarine Ecosystem with Special
Reference to Sundarbans, India

187

4.2 Biological analysis

4.2.1 Phytoplankton biomass (Chlorophyll-a)

Chlorophyll samples were filtered through Whatman GF/F (0. 45 µ) filters and extracted in
acetone in dark and refrigerated condition. Chlorophyll-a was determined
spectrofluorimetrically (Ventrick & Hayward, 1984).

4.2.2 Phytoplankton cell density

Direct estimation of phytoplankton cell abundance and diversity was performed by cell
counting method. Surface phytoplankton was collected and the Lugol's preserved
subsamples (1-2 liter) were used for quantitative enumeration utilizing a Sedgwick-Rafter
counting chamber and Zeiss research microscope according to UNESCO PROTOCOL
(1978). Several keys and illustration were consulted to confirm identification (Perry, 2003;
Tomas, 1997).

4.2.3 Zooplankton abundance and diversity

Zooplankton samples were collected monthly basis both in night and day time for
comparative and quantitative assay. Collection was performed both vertically and
horizontally. A long Bongo net, with mesh size 150 µm was used to collect sample. The
volume of water flowing through the net was measured by a digital flow meter (Model no. -
2030R, General oceanics).

4.2.4 Bacterioplankton abundance

Fluorescence microscope was used to estimate the total number of bacteria. Immediately
after sampling, 50 ml of seawater was preserved with 25% gluteraldehyde (0. 2-µm-
prefiltered) and stored in cold dark environment to prevent reduction of counts. Cells of
bacteria were collected onto a 25-mm black polycarbonate nucleopore membrane with a 0.
2 µm pore size and stained with acridine orange. At least twenty random fields were
counted in a Zeiss confocal fluorescence microscope coupled with an image analysis
system (Bianchi & Giuliano, 1995; Hobbie et. al., 1976). Viable count of Bacterial colonies
was also performed using Luria-Bertani medium by serial dilution method (Cappuccino &
Sherman, 2007).

4.2.5 Primary productivity

Primary productivity of water was measured by light and dark bottle method according to
the guidance of APHA(1998). Samples collected from each pre selected depth (on the basis
of light availability) were taken in triplicate light bottles. Dissolved oxygen of the initial
bottles was fixed with NaI-NaOH and $MnCl_2$ in the beginning of incubation period. At the
end of the incubation period light and dark bottles were similarly fixed and all the bottles
were brought back to laboratory in cold condition for analysis. Then dissolved oxygen
concentrations were estimated by Winkler's method (JGFOS Protocol, 1994). NPP (net
primary productivity), CR (community respiration) and GPP (gross primary productivity)
can be estimated using the following equations

- NPP = Light bottle – Initial bottle
- CR = Initial bottle – Dark bottle with nitrification inhibitor.
- GPP = Net primary productivity + community respiration

4.3 Statistical analysis

The results were expressed as differences between the groups considered significant at p < 0. 05. Data comparison and influence of the environmental factors on phytoplankton were evaluated by stepwise multiple regression (Manna et. al., 2010) Different statistical analysis and correlation regression analysis were performed using the software STATISTICA.

5. Results and discussions

5.1 Hydrodynamic parameters and their significance

Basic objective of hydrodynamic study is to understand the processes active within the tidal estuaries of Sundarbans; to develop methods so as to quantify the relative importance of river inflow discretely with respect to other forcing parameters like wind, tide, earth's movement etc as a dynamic forcing; to understand the relative importance of flow pattern as well as quantum of water coming as an input/output from different channels (creeks). Water movement and the consequent distributions of nutrients and other chemicals in an estuary are dependent on the hydrodynamic condition of the estuary. A hydrodynamic model is therefore a tool to understand the distribution pattern and availability of different nutrients within the estuary. This model is dependent on many physical parameters like tidal discharge, wave and meteorological forcing. In an estuarine condition, at the seaward boundary, tidal forcing drives the model. Tidal regime in an estuary ultimately determines the amount of sea water to be pushed into the estuary from the open sea carrying sediments as well as nutrients. Thus the boundary tide is most important and is usually specified by a water level time series, a velocity time series or a set of tidal harmonics. All these data need to be acquired from a set of water level measurements using different kinds of tide gauge and current meters. Unfortunately, for Sundarbans, such work has yet to be carried out. The freshwater discharges from rivers at the uplands also play an important role.

Sundarbans can be characterized with six North – South bound estuaries namely, Saptamukhi, Thankuran, Matla, Bidya, Goasaba and Raimongal. Each of these estuaries is different from each other in hydrodynamic set up. Except Raimongal, none of these estuaries are having freshwater discharge at head ward portion. These estuaries are more than 70 kilometers in length starting from its head ward point to the meeting point at the Bay of Bengal. A considerable portion of these estuaries pass through the inhabited islands, from where the supply of nutrients is in the form of agricultural field wash and aquaculture pond effluents and hardly any mangrove detritus is available. Thus, the availability of nutrients in these estuaries depends on the mixing within the estuaries, which in turn depends on tides and current.

Perhaps the most important data required is for tidal and surge information, however such data are almost entirely absent at the moment (Bhattacharyya et al, 2011). Delta Project report of 1968 (Delft, 1968) may be considered as the only published account of tidal range in Sundarbans. One permanent tide gauge station is located at Sagar Island in the Hooghly

Estuary and, although records have been kept here since 1937 these are a made using visual staff record operating only during daylight hours and are therefore incomplete. In order to overcome this problem, automatic tide gauge (Valeport make) instruments were placed within several estuaries of Sundarbans which indicated certain interesting hydrodynamic condition.

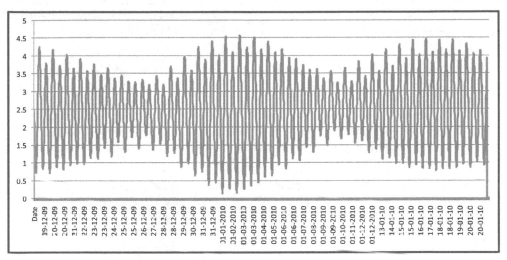

Fig. 2. Tidal fluctuation in Saptamuki mouth near Sitarampur

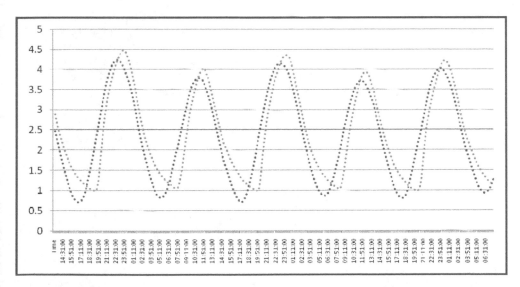

Fig. 3. Saptamukhi tides (in meters) December 2009 Red graph indicates tide at the mouth of Saptamukhi while blue colour indicates tidal fluctuation at Milon More about 50 kilometers away from the Saptamukhi mouth

It was observed that tidal range along the Sundarbans estuaries vary from place to place within the estuary. On an average, tidal fluctuation in Sundarbans estuary is around 5 meters depending on the lunar cycle, as is evident from the Figure. 2. But, tidal fluctuation near the mouth is comparatively smaller in range. As the tidal current pushes within the estuary, tidal range increases due to funneling effect of estuary. In case of Sundarbans, after travelling a certain distance tidal fluctuation starts receding due to bed resistance and ultimately dies down. It is quite interesting to observe the tidal regime along the estuary since ultimately this determines the availability of nutrients within the estuary.

During the period of the present study two automatic tide gauges were simultaneously deployed in the Saptamukhi estuary – one near its mouth (where the width of the estuary is about 1. 3 kilometers) with Bay of Bengal and the other at the head ward position at a place near Milan More (where the width of estuary is only 60 meters), at a distance of about 50 kilometer northwards from the first tide gauge. Incidentally, Saptamukhi is the westernmost estuary in Sundarbans and a number of mangrove islands including Lothian Island (which is a wildlife sanctuary) are located within it. The variance in tidal behavior pattern is quite obvious (Fig. 3).

It was interesting to note, that near the mouth tide is essentially symmetric in nature indicating that the time taken by high tide to reach the climax is exactly the same as that of low tide. Hence there is no additional residence time of tidal water within the estuary. However, at the Milon More, the tide is asymmetric in nature indicating that time taken by high tide to reach climax is much less compared to that of draining out of tidal water during ebb tide. Current speed within Sundarbans was found to be varying between 140 to 180 cm/sec. The current direction also is also controlled by the geomorphology of the creeks.

The Sundarbans estuary is thus a flood dominated estuary. The nutrient rich tidal water has a more residence time within estuary which is being used by the phytoplankton, the primary producer of this ecosystem. Thus the hydrodynamic set up helps in making this estuary so productive. It should be mentioned that the mangrove islands are all sea facing and water present within the creeks and estuary around the mangrove forest are always rich in nutrient, which is pushed well inside the estuary through tidal water.

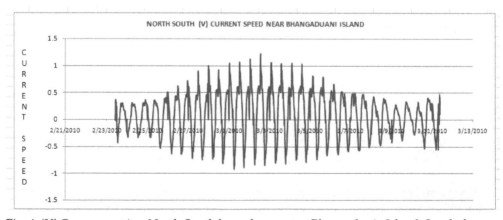

Fig. 4. 'V' Component i. e. North-South bound current at Bhangadunia Island, Sundarbans

Interplay of Physical, Chemical and Biological Components in Estuarine Ecosystem with Special
Reference to Sundarbans, India

191

During the period of study, several current meters were deployed in different estuaries of
Sunderbans to measure the current speed and velocity of tidal current for a period of one
month. This is for the first time such an exercise was carried out for Sundarbans. It was
observed that the tidal current velocity always depends on the lunar cycle and the phase of
tide, which is identical in case of all estuaries.

The U-V component of current the tidal current was then calculated at each spot to assess
the North-South current and East-West current. Interesting in all places it was found that the
V-component of current is the predominant current over the east-west current. Figure. 4 and
Figure. 5 indicate the comparative nature of these two components. The major estuaries of
Sundarbans are having a north-south trend and flanked by embankments along east-west
sides. Thus, observed current patterns indicate that whatever nutrients are discharged by
the mangrove islands along the southern boundary of Sundarbans are carried deep inside
the estuaries due to current pattern.

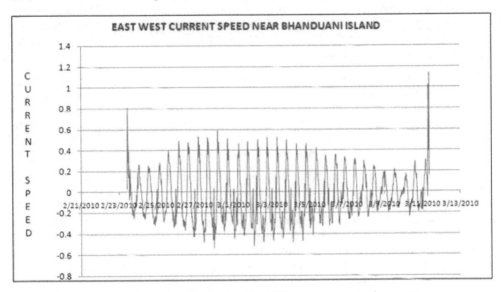

Fig. 5. 'U' component of current velocity i. e. East-West component of currentnear
Bhangadunia Island, Sundarbans.

Thus the hydrodynamic studies along estuaries of Sundarbans indicate that the
hydrodynamics play the major role for distribution of nutrients along the estuaries of
Sunderbans from the southern tip. While current is the driving force for transport of
nutrients to the distant parts of the estuaries, tidal regime ensures the availability of the
nutrients within the estuary for a longer duration to make it more productive.

5.2 Physicochemical characteristics

Physicochemical characters of estuarine ecosystem mainly depends on three factors, namely
dissolved oxygen concentration, salinity and sedimentation load within the water body.
Physical factors like temperature, pH, turbidity, salinity and dissolved oxygen in a water body

vary quickly relative to biological and chemical transformations. In general, the estuarine environment is oxidizing near the sediment–water interface and more reduced deeper in the sediment. The estuarine circulation movements are the primary mechanism to change the distribution profile of dissolved material in time and space in fresh and ocean water.

5.3 Temperature, pH, turbidity, salinity and dissolved oxygen

Sundarbans are located on shores in the tropics and enjoys tropical monsoon type of climates. The average rainfall of the region amounts to approximately 1750-1800 mm. summer and winter temperature varies between 26^0C-40^0C . The temperature of the surface water varied between 21^0 C-35^0C and significant variation is observed in temperature of different locations in the estuary.

Sundarbans estuarine pH levels generally average from 7. 0 - 7. 5 in the fresher sections, between 8. 0 - 8. 6 in the saline areas. Slightly alkaline pH of seawater is due to the natural buffering from carbonate and bicarbonate dissolved in the water (Volunteer Estuary Monitoring, 2006). The pH values of the Sundarbans estuary were slightly basic and remained almost constant (7. 9-8. 2) except during the monsoon months when a slight but insignificant decrease was noticed.

Turbidity indicates water clarity and it is the measured by light scattering by suspended particles in the water column. Several factors are responsible for water turbidity; suspended soil particles (including clay, silt, and sand); tiny floating organisms (e. g., phytoplankton, zooplankton, and bacterioplankton); and small fragments of dead plants (Voluntary Estuary Monitoring Manual, 2006). In Sundarbans estuary turbidity ranges from 35-150 NTU, and highest index being observed in monsoon.

In estuaries mixing of sea water with fresh water causes brackish water to be more saline than fresh water but less saline than sea water. Salinity of estuaries usually increases away from a freshwater source such as a river, although evaporation sometimes causes the salinity at the head of an estuary to exceed seawater. Vertical salinity structure and nature of salinity variation along an estuary is the unique feature of coastal water ways (Santoroet. al. 1989). Sundarbans estuary situated in the delta of Bay of Bengal showed salinity gradient from the upstream to the downstream part and also margin to central part (Baidya, & Choudhury, 1984). In Sundabans estuary salinity ranges from 11-25 PSU, being highest in dry season and lowest in wet season (Manna et. al., 2010). In Sundarbans estuary tidal action is very strong and practically it is the only regulatory factor, thus water is well mixed from top to bottom and the salinity approaches that of the open sea.

5.4 Dissolved oxygen (DO)

DO is one of the most important controlling factor regulating presence of estuarine species. In addition to support respiration, oxygen is needed for decomposition, an integral part of an estuary's ecological cycle is the breakdown of organic matter (Volunteer Estuary Monitoring Mannual, 2006). DO concentration in the water column is highly dependent on temperature, salinity and biological activity. Tidal estuaries are generally characterized by high DO level. Dissolved oxygen concentration was steady all along the stretch over Sundarban estuary varying between 3. 3 to 9. 5 mg/L with comparatively higher values during November-February (bloom season).

5.5 Nutrients

Autotrophic nutrients are important for estuarine ecosystems and are essential for
sustenance of the marine ecosystems because they are the raw materials for primary
producers. Main estuarine nutrients include phosphate, nitrate, nitrite, ammonia and
silicate. Nitrogen, phosphorus and silica are the key nutrients that generally limit
phytoplankton growth in natural waters. Silicate is a primary growth limiting nutrient for
diatoms. It is reported that N-limitation is a wide spread phenomenon in tropical lakes
rivers and estuaries, e. g., Mandovi-Zuari (Ram et al., 2003), Cochin estuary (Gupta et. al.,
2009) and Hoogly estuary (Mukhopadhyay et al., 2006). Sundarbans estuary was
phosphorus limited in postmonsoon and nitrogen-limited in premonsoon and monsoon.
However, seasonal phosphorous limitation characteristic was found in several estuaries (eg.
Gle et al. 2008; Xu et al. 2008). Sundarbans estuary is a nutient rich tropical estuary with
high nutrient influx, where a huge quantity of leaf litter is loaded to the estuarine water
from adjacent mangrove forests. Besides, land mass wash off during monsoon and drainage
waste from shrimp culture farms also contributed to this huge nutrient load. In Sundarban
estuary the phosphate concentration varied from 0. 4 to1. 0 µmol/L, the nitrate
concentration varied from 2. 4-39. 9 µmol/L, nitrite concentration ranged from 0. 6-1. 6
µmol/L and silicate concentration varied from 4. 8 to 49. 1 µmol/L.

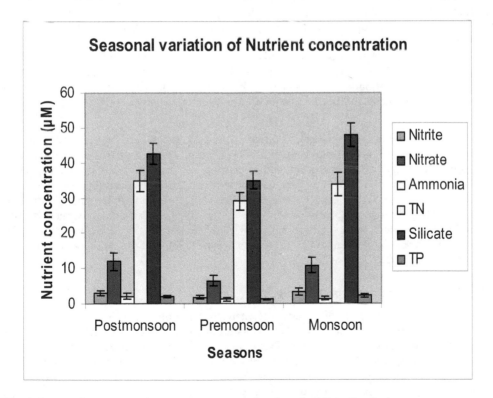

Fig. 6. Seasonal variation of nutrient concentration (µmole L[-1]) in Sundarban estuary.

6. Biological components

6.1 Autotrophic nutrition and food web

In an estuary, the plants and other primary producers (algae) convert energy into living biological materials (Ryther, 1969). Detritus feeders, plant grazers, and zooplankton are the primary consumers, and the secondary consumers and tertiary consumers include estuarine birds, ducks, invertebrate predators, and fish. Excreta and detritus pass to the decomposer tropic level where microorganisms break down the material. At each stage in this trophic sequence matter and energy are consumed, and some of it is excreted as waste, or converted into body growth or heat after respiration (McLusky and Elliott, 2004).

Plankton is one of the important components of any aquatic ecosystem among which phytoplanktons are the primary source of food in the marine pelagic environment, initiating the food-chain which may culminate even in large mammals (Waniek and Holliday, 2006). Studies of phytoplankton are essential to understand food chain dynamics in aquatic ecosystems (Sieburth and Davis, 1982). In oligotrophic waters, the base of the food chain is composed of very small cells in the size range 0. 2–2 µm ['picoplankton' (Sieburth et al., 1978)] and the microbial loop dominates the pelagic food web (Fenchel, 1988). Conversely, when an import of mineral nutrients takes place, the base of the food chain comprises larger phytoplankton (diatoms and dinoflagellates), which are more readily eaten by zooplanktons; under these conditions, the classic pelagic food chain dominates (Fenchel, 1988). Autotrophic picoplankton ['picophytoplankton' (Fogg, 1986)] is represented by prokaryotic coccoid cyanobacteria, frequently of the genus *Synechococcus*, prochlorophytes (Chisholm *et al.*, 1988) and small eukaryotic cells (Johnson and Sieburth, 1982). Picophytoplankton is important contributors to total phytoplankton biomass and primary production in all aquatic environments (Stockner, 1988). They dominate the total phytoplankton biomass and production in oligotrophic environments (Fogg, 1986; Weisse, 1993).

Phytoplankton biomass and primary production mainly depend on nutrient dynamics of coastal and estuarine ecosystems (Nixon, 1995; Cloern, 1999). Estuarine phytoplankton production is mostly dependent on either nutrient or light availability (Riley, 1967; Williams, 1972; Fisher et al, 1982). In case of nutrient rich turbid estuaries light is the major controlling factor and restricted light availability may alter phytoplankton production (Wofsy, 1983; Pennock, 1985; Harris, 1978;Falkwski, 1980). Estuarine dynamics is well studied in temperate system such as Chesapeake Bay (Bonyton et. al. 1982; Hording, 1994; Kemp et. al., 2005), San Francisco Bay (Review: Cloern, 1996), and the Baltic sea (Graneli et. al., 1990; Conley, 2000).

6.2 Heterotrophic bacterial production

The production and structure of aquatic ecosystems depend on interactions between energy mobilizers, i. e., phytoplankton and bacterioplankton (Jones, 1992) and the abiotic factors that control their activity (Jones, 1998). Bacteria can vary from being consumers of energy released by phytoplankton, to be independent mobilizers of energy to the ecosystem, using carbon compounds imported from the catchment as a source (allochthonous carbon) of energy (Jones, 1992) Consequently, the role of bacteria as an energy mobilizer for the pelagic ecosystem increases with increasing input of allochthonous DOC (Jansson et. al., 2000).

Interplay of Physical, Chemical and Biological Components in Estuarine Ecosystem with Special
Reference to Sundarbans, India

195

Therefore bacterioplankton may be regarded as a highly important component of the pelagic ecosystem. In case of mangrove dominated Sunderbans estuary where a huge quantity of leaf litters is loaded to the adjacent estuarine water nitrogen may be the main currency in determining overall productivity and heterotrophic bacterial production may exceed phytoplankton primary production.

Many different factors may control the heterotrophic bacterial activity of aquatic ecosystem namely, temperature (Oachs et. al., 1995; Rae & Vincent; 1998), inorganic nutrients (Jansson et. al., 1996; Kroer, 1993), dissolved organic carbon (Tranvik, 1988; Hobbie et. al., 1996) etc. Organic carbon metabolism of clear water bodies like lakes usually dominated by autochthonous sources (Baron et. al., 1991) because of low export of organic matter from surrounding soils whereas in case of nutrient rich estuaries and bays receiving high loading of allochthonous DOC, the bacteria may be relieved of their close dependency on phytoplankton carbon. This may cause bacteria growth to be limited by nutrients (Jansson et. al, 1996; Kroer 1993). Temperature also affects the bacterial growth rate (Oachs et. al., 1995; Rae & Vincent 1998, Tulonen et. al., 1994). In case of nutrient rich Sunderbans estuary bacterial population showed an exponential relationship with temperature (Manna et. al., 2010).

6.3 Community respiration and nitrification

In a water body all the living organisms respire and consume oxygen (O_2). Respiration provides a simple and straightforward measure of heterotrophic activity that can be directly related to the oxidation of organic matter (Williams, 1981; Hopkinson et al., 1989 & Biddanda et al., 1994) and it is regarded as a key index of the energy used by consumers at a given time and place (Biddanda et. al., 1994 & Pomeroy et. al., 1968).

Episodic oxygen depletion in the water column is a common feature in many coastal and estuarine areas during summer (Turner & Rabalais, 1991; Kemp et al., 1992). In a variety of estuaries nitrification is a major contributor to total pelagic oxygen consumption(Kausch, 1990; Pakulsht, 1995) and records of nitrification rate in addition to primary production and respiration can be an important parameter to understand variations in oxygen concentration in the water column. However, few measurements of the actual rates of pelagic nitrification have been reported from coastal and estuarine waters and studies on factors regulating marine nitrification are still scant in literature (Owens, 1986; Berounsky & Nixon, 1993).

Sunderbans estuary is designated as a moderately productive estuary with an annual integrated phytoplankton production rate of 2. 9 -5. 4 µgC/L/hr and community respiration 1. 75-3. 2 µgC/L/hr (Unpublished data).

Seasons	GPP(µgC/L/hr)	CR (µgC/L/hr)
Postmonsoon	4. 5-5. 4	1. 75-1. 95
Premonsoon	4. 2-4. 4	2. 2-3. 2
Monsoon	2. 9-3. 5	1. 8-2. 07

Table 1. Seasonal variation of primary productivity in Sundarban estuary

6.4 Processes associated with microorganisms

Cycle of energy and matter in estuaries is closely related with microbial activity, half of the aerobic and anaerobic transformations of organic matter in salt marsh are the result of microbial metabolism. Chemical transformation mediated by marine microbes play a critical role in global biogeochemical cycles. Coastal regions show the highest concentration of nutrients and microorganisms and the least light penetration whereas the open ocean is largely olegotrophic (extremely low concentration of nutrients and microorganisms). The concentration of heterotrophic microorganisms determines the Biochemical Oxygen Demand (BOD) i. e. the amount of oxygen removed from water by aerobic respiration. So the coastal region has a high BOD compared to open ocean. A huge energy cycle is carried out by the photosynthetic microbes living on the ocean surface. These microscopic communities are responsible for 98% of primary production (Whitman et. al., 1998, Atlas & Bartha, 1993). This makes ocean microbes one of the major sinks of atmospheric carbon dioxide, a process termed "carbon sequestration". They also mediate all the biogeochemical cycles in the oceans (Atlas & Bartha, 1993).

6.5 Carbon and nitrogen cycles

Bacteria show a variety of metabolic pathways related to carbon flow and cycling. Carbon fixing rate of phytoplankton shows marked seasonal fluctuations in hydrographic and nutrient parameters. As many of the sediment and water-logged soils of estuaries are anoxic, anaerobic decomposition is important. Complex organic matter is used by the fermenters and dissimilatory nitrogenous oxide reducers. The sulfate reducers and methane producers were once thought to have more restricted distributions (John et. al., 1989). Researchers suggested seasonal and inter annual dynamics of free-living bacterioplankton and labile organic carbon available to microbes along the salinity gradient of estuaries. Bacterioplankton abundance may be an important indicator of ecosystem health in eutrophied estuaries, because of the positive relationships between bacterioplankton abundance, microbially labile organic carbon (MLOC), and dissolved oxygen (Leila et. al., 2007).

Nitrogen is a major limiting nutrient for primary production in estuaries. The N-cycling processes that are dominated by microbial activity include nitrification, dissimilatory nitrous oxide reduction, and nitrogen fixation. Nitrogen cycling in estuaries is related to the water mixing and microbial community dynamics.

7. Diversity and distribution of estuarine organisms

7.1 Phytoplankton – The primary producer

Algae belong to a highly diverse group of photoautotrophic organisms with chlorophyll a and unicellular reproductive structures, which are important for aquatic habitats (Ariyadej et. al., 2004) Phytoplankton species composition, richness, population density, and primary productivity vary from coast to coast and sea to sea depending upon the varying hydro biological feature. (Prabhahar et. al., 2011) and seasonal and spatial distribution of plankton in the estuary were discernable). Changes in species composition and dominance of phytoplankton can be mediated by a variety of mechanisms including ambient temperature, light penetration, nutrient supply, and removal by zooplankton (Reynolds 1993).

Interplay of Physical, Chemical and Biological Components in Estuarine Ecosystem with Special
Reference to Sundarbans, India

197

Sundarbans estuary is formed by a complex network of upstream rivers where the spatial and seasonal variations of some hydrochemical characters are quite prominent and water quality parameters of rivers showed marked variation in different seasons which greatly influence species composition and quantitative abundance of planktons. In Sunderbans phytoplankton abundance ranged from 7. 25 X 10^4 cells / l (June) to 9. 8 x10^6 cells /L (February) (Manna et al., 2010).

Season (Period)	Planktonic abundance (Cells/L)	Biomass (Chlorophyll-a concentration, µg/L)	Dominant taxa
Postmonsoon (Nov – Feb)	9. 25 x 105 – 9. 8 x 106	19. 9 - 36. 5	*Coscinodiscus sp.*, **Chaetoceros sp.**, *Baccteriastrum sp., Thalassiosira sp., Planktoniella sp., Triceratium sp.*
Premonsoon (Mar– Jun)	7. 25 x 104 - 2. 8 x 105	10. 5 - 28. 7	*Navicula sp., Thalassionema sp., Synedra sp. Diatoma sp., Nitzschia sp, Protoperidinium sp., Chlorella sp.* **Dunaliella sp.**
Monsoon (Jul – Oct)	2. 44 x 105 -5. 2 x 106	6. 9 - 22. 7	*Navicula sp., Thalassionema sp, Cosmarium sp., Closterium sp., Oscillatoria sp. Stigonema sp Pyrocystis sp., Anabena sp* **Tricodesmium sp.**

Table 2. Distribution and Abundance of Phytoplankton in Sundarban estuary

Sunderbans has a highly diverse algal flora comprised of both benthic and planktonic forms ranging from the freshwater to marine environments (Gopal and Chauhan, 2006) and showed the highest species diversity in all seasons. Noticeable variation of phytoplankton forms was also observed in seasons and sampling locations due to variations of water quality parameters, like pH, salinity, TSS and nutrients and DO. Phytoplankton community was observed to be dominated by diatoms (Biacillariophyceae) followed by Pyrrophyceae (Dinoflagellates) and Chlorophyceae and higest abundance was noticed in postmonsoon (Biswas et al. 2010; Manna et al. 2010) Centric Diatoms predominated in winter months and Pennates in summer whereas Chlorophyceae, Cyanohyceae and Euglenoids dominated the estuary in monsoon. During premonsoon the dominant phytoplankton were species of Ditylum, Ceratium, Biddulphia, Chaetoceros, Coscinodiscus, Thalassiothrix, Rhizosolenia Nitzschia and Thalassionema. However, during postmonsoon phytoplankton species of Bacteriastrum, Biddulphia, Protoperidinium and Ceratium were most dominant (Fig 6a) But in monsoon species of Skeletonema, Fragillaria and some blue green algae, green algae and also euglenoids are quite common. The average phytoplankton load is higher mostly in postmonsoon (Manna et al. 2010). The eastern part of the estuary is dominated by phytoplankton species like Biddulphia diatoms and green and blue green algae, while the central part is dominated by a variety of diatom species viz, Chaetoceros, Coscinodiscus, Bacterioastrum, Cyclotella, Ditylum, Skeletonema, Thallassiothrix, Thalassionema and Triceratium. In contrary, the western region is dominantly represented by species of

Fragillaria, Gyrosigma, Nitzschlia and Bacillaria. The seasonal variation of phytoplankton load indicates that there is a bimodal pattern of distribution with one in premonsoon (May) and other in postmonsoon (November) (Bandyopadhyay, 2003 :Biwas et al. 2010; Manna et al., 2010). Such biomodal seasonal cycle is a typical feature throughout the coastlines in India. (Mani, 1986; Gopinath, 1972; Gouda. 1991) The phytoplankton taxa of Sunderban estuary in general resembles with those of coastal waters, estuarine and near shore region of Goa (Devassy. & Goes, 1988) Cochin back waters, (Devary, & Bhattathiri, 1974) Hooghly river, (Santra et al. 1989; Santra & Pal; 1989), Rushikulya estuary (Gouda, R. & Panigrahy, 1989), Bahuda estuary, (Mishra & Panigrahy, 1995) and Porto Novo (Kannan & Vasantha, 1992). Only difference lies in the occurence of bloom forming diatoms. At high salinity level (25 PSU) halo- tolerant phytoplankton species like Dunaliella salina, Chlorella marina, C. salina, Grammatophora marina, Fragillaria oceanica dominated the estuary which are definitely better adapted to the high saline environment. Bio-indicator species like Polykrikos schwartzil, Dinophysis norvegica and Prorocentrum concavum points to moderately polluted water quality of the Sunderbans estuary (Manna et al., 2010).

Class	Genus	Species
Bacillariophyceae	61	151
Pyrrophyceae	15	53
Cyanophyceae	21	23
Chlorophyceae	34	44
Chrysophyceae	3	8
Euglenophyceae	3	5
Dictyophyceae	3	2
Haptophyceae	1	2
Xanthophyceae	1	1

Table 3. Composition of Phytoplankton Classes.

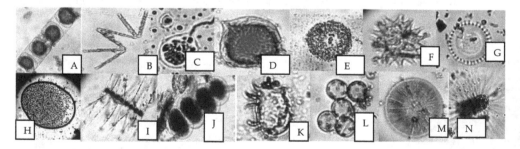

A) *Stephanopyxis palmariana* B) *Thalassionema costatum* C) *Trachelomonas volvocinopsis* D) *Protoperidinium diabolus* E) *Phaeocystis globosa* F) *Pediastrum duplex* G) *Cyclotella stelligera* H) *Aphanothece smithii* I) *Chaetoceros impressus* J) *Peridiniella catenella* H) *Gonyaulax spinifera* L) *Gonium pectorale* M) *Cyclotella striata* N) *Mallomonas tonsulata*

Fig. 7a. Representative phytoplankton taxa of Sundarbans estuary

Interplay of Physical, Chemical and Biological Components in Estuarine Ecosystem with Special
Reference to Sundarbans, India

199

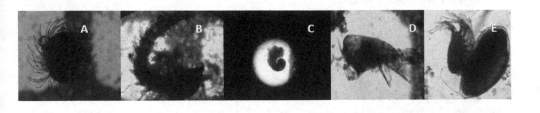

A. *Sabellaria cementarium* B. *Pagurus longicarpus* C. *Gastropod veliger* D. *Corycaeus amizonicus* E. *Cirriped-Cypris* F. *Lucifer sp.* G. *Acrocalanus gracilis* H. *Cyphonautes larva* I. *Nematode larva* J. *Fish larva*

Fig. 7b. Representative zooplankton taxa of Sundarbans estuary

7.2 Zooplankton – Primary consumer

Zooplankton are organism that belonging the secondary level in food web and plays major role from producers level to upper most consumer level. They are more abundant within mangrove water-ways than in adjacent coastal waters (Robertson and Blabber 1992). Zooplankton support many major fisheries and mediate fluxes of nutrients and chemical elements essential to life on Earth.

Copepod dominated Zooplankton abundance in Sunderbans estuary (Fig. 6b) forming 40%-65% of total composition except three months. During February, April and September the dominant species were Tintinnopsis cylindrica, Polychaete larvae(all types)and shrimp larvae respectively. Around 52 taxa of nine major phyla and two minor phyla were identified during the study period. The total counting ranged 1. $92×10^2$-7. $6×10^3$ cells L^{-1}. Monthly average abundance of Zooplankton ranges indicated the health status of the Sundarbans estuarine aquatic ecosystem. The nutrient upload and water mass upwelling changes the whole pattern of species richness throughout all seasons . Species Richness was highest in Monsoon than the other two seasons where as diversity index was lowest in Post monsoon. Mollusc larvae were dominated in some areas. Mysids and Chaetognath were largely found where as Copepod nauplius, Balanus nauplius, Gastropod veliger and Crab Zoea are more or less same in number in all the stations including Lucifer sp., Protozoans and Bryozoan are distinct Zooplankton which were also found. Nematode larva was a complicated one which only found in estuary but also less in number. In all Estuarine system of World Copepods are the dominating one with respect to other taxa but in case of Indian Sundarbans it fluctuates due to species dominancy in particular time with respect to some cause like nutrient availability or any other environmental factors.

7.3 Fungi – The heterotrophs

The number of fungi living in estuaries is extremely large. Some of fungi are unique in estuaries, while others have a broader range of habitats. Aquatic fungi and yeast dominate species in aquatic environment, few of fungi associate with particles or solid matters in the water. In sediments, the active species of fungi primarily are found in surface aerobic zones. The densities of fungi decrease rapidly with soil depth, but the spores of fungi are found throughout sediments (John et al., 1989). Leaf inhabiting fungi of mangrove plants are known. *Khuskia oryzae* has been reported for the first time from India, among seven species of fungi that exist on mangrove leaf surface of Sunderbanss. There are two new parasitic fungi namely *Pestalotiopsis agallochae* sp. and *Cladosporiummarinum* sp. existing on infected leaves of *Excoecaria agallocha* and *Avicenniamarina* (Pal and Purkayastha, 1992a). Sixteen fungi are isolated from leaves of mangrove plants of Sunderbanss, West Bengal and their growth response to tannin; extracellular pectolytic enzyme (PE) activity and degree of inactivation of PE due to presence of tannin are tested in vitro.

7.4 Bacterioplankton – The detrivore

Microorganisms constitute a huge and almost unexplained reservoir of resources. Microorganisms are the richest repertoire of molecular and chemical diversity in nature. They underlie basic ecosystem processes such as the biogeochemical cycles, food chains and maintain vital relationships between themselves and higher organisms. The enormous number of microbes, their vast metabolic diversity and the accumulation of mutations during the past 3.5 billion years should have led to very high levels of genetic and phenotypic variation (Sogin et. al., 2006).

The world's oceans are teeming with microscopic life forms. Normal microbial cell counts are greater than 10^5 cells per ml in surface sea water (Porter & Feig, 1980). It implies that the oceans harbour 3.6 x 10^{29} microbial cells with a total cellular carbon content of approximately $3x10^{17}$g (Whitman et. al., 1998). In contrast, the total microbial cell number of deep ocean waters (>1000 m) is only $8x10^3$ – $9x10^4$ (Karner et. al., 2001 & Cowen, 2003), half of which are archae (Santelli et. al., 2008). Net primary productivity in the global ocean is estimated to fix 45-50 billion tons of carbon dioxide per year (Falkowski et. al., 1998).

Bacteria are the most abundant organisms in the estuary, averaging between 10^6 to 10^7 per ml organisms in water and 10^8 to 10^{10} per dry weight of sediment. Sediments and salt marsh soil generally harbor more bacteria per unit volume than does the water column. Within the water column, high densities may be found in the surface layer than subsurface layer. Aerobic and facultative anaerobic bacteria are most common, and *Pseudomonads* and *Vibrio* are the most often isolated species. Sediment and waterlogged soils show very high densities of bacteria, which decrease in abundant with depth of soils. Higher bacteria densities have been found in most estuaries than in nearby coastal seawater and river water (John et al., 1989). In Sundarban estuary bacterial population ranged from $1.62X10^3$ – 3. $18X10^6$, lowest in January (3.68×10^6 CFU L[-1]) and the highest in May (8.9×10^8 CFU L[-1]) (Manna et. al., 2010).

Marine environment is dominated by microscopic protists and prokaryotes. However, it is widely accepted that current and traditional culture based techniques are inadequate to study microbial diversity from environmental samples. Our understanding of marine

Interplay of Physical, Chemical and Biological Components in Estuarine Ecosystem with Special
Reference to Sundarbans, India

201

microbial communities has increased enormously over the past two decades as result of culture independent phylogenetic studies. Recent advances in molecular techniques are adequate to describe the microbial diversity in a marine sample based on 16S rRNA sequence diversity.

Period	Bacterial count	
	Plate count (CFU ml^{-1})	Fluorescence count (Cells ml^{-1})
Postmonsoon (Nov '08 - Feb '09)	$1.62 \times 10^3 - 4.12 \times 10^4$	$8.15 \times 10^4 - 1.65 \times 10^6$
Premonsoon (Mar '09 - Jun '09)	$3.65 \times 10^4 - 1.85 \times 10^6$	$8.57 \times 10^5 - 6.58 \times 10^7$
Monsoon (Jul '09 - Oct '09)	$6.13 \times 10^4 - 3.18 \times 10^6$	$2.07 \times 10^6 - 9.11 \times 10^7$

Table 4. Seasonal variation of bacterial abundance in Sundarban estuary

It is difficult to grow marine bacteria in culture. Only 0. 1- 10% bacteria are culturable in currently used culture media. Culture-dependent methods do not accurately reflect the actual bacterial community structure. Furthermore, (1) all techniques rely on cultivation are time consuming and expensive as are the physiological and biochemical differentiation tests; (2) after many generations necessary to form plate colonies, the organism may deviate from its physiology, and possibly even from genotypic mix, of the population in nature; (3) though many advances have been made in microbiological culture techniques, it is still not possible to grow a majority of bacterial species using the standard laboratory culturing techniques (Bakonyi et. al., 2003). Only a minor fraction (0. 1-10%) of the bacteria can be cultivated using standard techniques; (4) The biggest drawback in exploring bacterial biodiversity is that phenotypic methods can be applied only on bacteria which can be cultured and (4) it offers a very limited insight into the spatial distribution of the microorganisms (Pace et al., 1986; Holben & Tiedje, 1988; Ward et al., 1992; Amann et al., 1995). Microbial ecologists are turning increasingly to culture-independent methods of community analysis because of the inherent limitations of culture- based methods. Quantitative estimation of community composition can be inferred based on advanced fluorescence microscopic techniques using culture-independent methods. Useful

molecules for such studies include phospholipids, fatty acids and nucleic acids (Morgan & Winstanley, 1997) where as the microscopic techniques involve either the hybridization of fluorescent- labeled nucleic acid probes with total RNA extracted from water or hybridizations with cells in situ. Metagenomics, also known as environmental genomics, is the culture independent study of a community of microorganisms (Steele & Streit, 2005; Fieseler et. al., 2006). This relatively new technology provides a chance to study the other 99% of bacteria (Tringe & Rubin, 2005).

8. Water quality of the estuary

8.1 Nutrient influx and water quality

Estuarine water quality and habitat conditions are directly affected by fluxes of nutrients from the sediments, especially in summer when temperature is high and hypoxic and anoxic events typically occur. The magnitudes of these sediment oxygen and nutrient fluxes also appear to be directly influenced by nutrient and organic matter loading to the estuarine systems (Kemp and Boynton 1992). Both annual and interannual patterns demonstrate that when these external nutrient and organic matter loadings decrease, the cycle of organic matter deposition to the sediments, sediment oxygen demand, and the release of nutrients into the water column also decrease and water quality and habitat conditions improve (Boynton et al., 1995). The chemical parameters of water – primarily nitrates and phosphates – used in combination with productivity determine the trophic condition of estuaries as oligotrophic (low nutrient) through mesotrophic to eutrophic (high nutrient) (Bellinger and Sigee, 2011).

Study in Sundarbans estuary suggests that indicators of inorganic nutrients and plant productivity changed widely during the annual cycle of the estuary as shown below (Manna et. al., 2010; Biswas et al., 2010) (Table 7). The study indicates estuary remained eutrophic during winter and meso-eutrophic during monsoon and premonsoon (Manna et. al., 2010).

Parameter	Postmonsoon	Monsoon	Pre monsoon
Nutrient concentration (μmoll[-1])			
Total phosphorus	2-2.15	1.1-1.84	.1-1.98
DIN	21-25	6-22	5-15
Chlorophyll a concentration (μg l[-1])			
Mean concentration	28	19.5	17.5
Maximum concentration	40	28.7	22.7
Secchi depth (cm)			
Mean value	60	25	130
Minimum value	40	20	120

Table 5. Seasonal vaiation of physicochemical parameters in Sunderbans estuary

8.2 Eutrophication of the estuary

Eutrophication seems to be a global problem and nutrient enrichment is one of the most serious threats to near shore coastal ecosystems (Cloern 2001). The balance of water ecosystem is disturbed by eutrophication i. e. excessive fertilization, which, in turn, leads to increases in phytoplankton quantity and primary production. Estuaries receive considerable amounts of freshwater, nutrients, dissolved and particulate organic matter, suspended matter, andcontaminants from land and exchange materials and energy with the open ocean. Estuaries also receive nutrients and organic matter loads from wetlands such as marshes (Meybeck 1982; Kemp 1984, Howarth et al. 1996) and mangroves (Odum and Heald 1972; 1975; Twilley 1988; Robertson et al., 1992) The consequences of nutrient enrichment include algal blooms (Paerl 1997), coral reef degradation (Lapointe 1997& Hughes et al. 2003), loss of diversity and ecosystem resilience (Levine 1998 & Scheffe 2001) and, in

Interplay of Physical, Chemical and Biological Components in Estuarine Ecosystem with Special
Reference to Sundarbans, India

203

extreme cases, the development of "dead zones"(Rabalais et al. 2002). Eutrophication also initiates changes in phytoplankton community structure, decrease in diversity and frequency of harmful algal blooms. It can have significant deleterious effects on the beneficial uses of estuarine and marine waters.

The eutrophication of coastal waters is a problem of epidemic proportion and has disastrous short- and longterm consequences for water quality and resource utilization (Paerl, 1997; Nixon, 1995; Turner and Rabalais, 2003). Metrices based on phytoplankton quantity and productivity is widely used indicators of eutrophication in the status assessment of surface waters (HELCOM:2002; EEA 2007). There are a number of ways in which eutrophication of estuary manifest itself: increase in phytoplankton biomass (Harding & Perry, 1997) and macroalgae (Valiela et al., 1997) anoxia and hypoxia (Rosenberg 1990 & Kiirikki et al. 2006). The most commonly used indicator of eutrophication in waterbody, however, is chlorophyll a (Kauppila 2007). In Sundarbans estuary, chlorophyll-a concentration remained very high (>10 µg L^{-1}) most of the time throughout the year indicating the estuary was in eutrophic condition (Fig 7) (Jones & Fred Lee, 1982).

The Eutrophication may have detrimental effect on the mangrove vegetation. These negative consequences contrast with observations that marine plant growth, including that of tropical mangroves is enhanced with nutrient enrichment (Paerl 1997; Catherine et al. 2009) However, nutrient enrichment favours growth of shoots relative to roots (Grime; 1979, Lambers &, Poorter 1992 & Catherine et al. 2009) thus enhancing growth rates but increasing vulnerability to environmental stresses such as drought, that require large investment in roots for tolerance (Chapin 1991 & Catherine et al. 2009) Thus the benefits of increased mangrove growth in response to coastal eutrophication is offset by the costs of decreased resilience due to mortality during drought, with mortality increasing with soil water salinity along climatic gradient.

8.3 Phytoplanktons as an indicator of water quality

Phytoplankton is the most important producer of organic substances in the aquatic environment and the rate at which energy is stored up by these tiny organisms determine the basic primary productivity of the ecosystem. Phytoplankton satisfy conditions to qualify as suitable pollution indicators in that they are simple, capable of quantifying changes in water quality, applicable over large geographic areas and can also furnish data on background conditions and natural variability (Lee, 1999). Phytoplanktons are characterized for their rapid responses to alterations in environmental conditions (Reynolds 1984; Stolte et. al. 1994) such as anthropogenically introduced eutrophications of coastal waters (Richardson, 1997). The latter characteristic makes phytoplankton sensitive indicators of changes in aquatic ecosystems (Valdes-Weaver et. al., 2006). Their presence or absence from the community indicates changes in physio-chemical environment of the estuary (Rissik, 2009). More so, micro algal components respond rapidly to perturbations and are suitable bio-indicators of water condition which are beyond the tolerance of many other biota used for monitoring (Nwankwo and Akinsoji, 1992). Species diversity indices when correlated with physical and chemical parameters, provide one of the best ways to detect and evaluate the impact of pollution on aquatic communities (Maraglef, 1968) In India, Ganges-Brahmaputra estuary is particularly vulnerable to anthropogenic perturbations due to high nutrient loads from reverine discharge, increasing human population density and rapid economic growth (Seitzinger et. al., 2005; Mukhopadhyay et. al., 2006; Biswas et. al., 2010). Eutrophication as well as presence of toxic Dinoflagellates and Cyanophyceae in the tidal creek of Sunderbans estuary definitely revealed the deteriorated status of the water quality. (Manna et. al. 2010).

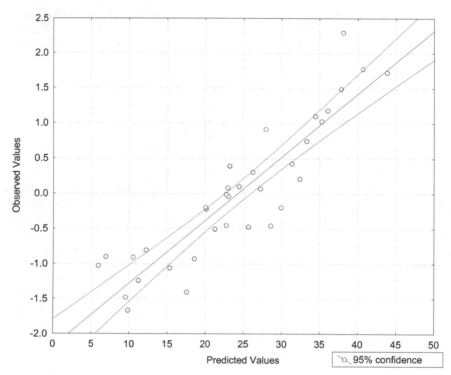

Fig. 8. Plot of observed versus predicted values of chlorophyll-a in Sundarbans estuary (Manna et al.; 2010)

Phytoplankton is one of the most rapid detectors of environmental changes due to their quick response to toxin and other chemicals. Pollution stess reduces the number of algal species but increases the number of individuals. A marked change in algal community severely affects species diversity (Biligrami, 1988). Eutrophication or organic pollution of aquatic ecosystems results in replacement of algal groups. It has been observed that many species are sensitive to nutritional loading, but equally good numbered are pollution tolerant. A numbers of reports are available on pollution –indicating and pollution tolerant algal species (Desei et. al. 2008). Similarly, a good number of indices have also been evolved to determine the trophic level of water ecosystems like Nygaard's algal index, Shannon and Weiner's species diversity indices and Palmer;s pollution index.

In general nutrient-deficient natural water harbouring low populations of algae, on addition of nutrients, increases the growth of the algae. The water appears dark green on excessive algal growth or the algal blooms. These algal blooms occur in highly enriched waters, especially that receiving sewage waste (Trivedi & Goel, 1984). Certain species of

Interplay of Physical, Chemical and Biological Components in Estuarine Ecosystem with Special
Reference to Sundarbans, India

205

phytoplankton grow luxuriantly in eutrophic waters while some species cannot tolerate waters that are contaminated with organic or chemical wastes. Some of the species indicate the clean waters are *Melosira islandica, Cyclotella ocellata* and *Dinobryon*. Pollution indicating plankton includes *Nitzschia palea, aeruginosa* and *Aphanizomenon flosaquae*. The latter two species have been found to produce toxic blooms and anoxic conditions. Some algae were found to cause noxious bloom in polluted water that tastes bad with intolerable odour.

Bioindicators are defined as species or communities which by their presence provide informations on the surrounding physical and chemical environments in a particulary site (Belliger & Sigee 2011) A good indicator species has characteristics like narrow ecological range, wide ranging distribution, easily identifiable characteristics, quick response to environmental stress and well defined taxonomy. The organic pollution influence algal flora more intensely than most other factors including DO, pH, light intensity, hardness of water and other type of pollutants (Palmer, 1969).

Genus	Family	According to Palmer	Sundarbans Estuary
		Pollution index	Pollution index
Euglena	Eu	5	5
Oscillatoria	Cy	5	5
Chlamydomonas	Ch	4	4
Scenedesmus	Ch	4	0
Chlorella	Ch	3	3
Nitzschia	Ba	3	3
Navicula	Ba	3	3
Stigeoclonium	Ch	2	2
Synedra	Ba	2	6
Ankistrodesmus	Ch	2	2

Ba-Bacillariophyceae; Cy-Cyanophyceae; Eu-Euglenophyceae: Ch-Chlorophyceae.

Table 6. Palmer's family pollution index (1969) with input from Sundarbans estuary

The algal species are rated on a scale of 1 to 6 (intolerant to tolerant) and index is derived by summing up the scores of all taxa that are present in sample. In computing Palmer's index an alga is considered to be present if there are 50 or more individuals per litre of sample.

Thus, the Palmer's index for Sunderbanss comes to very high (**33**). As any value higher than **20** is condidered to be high in Palmer's index, it tallies quite well with the result of trophic analysis as per OECD norms. Thus, there can be no second opinion about the fact that the estuary is eutrophic for the most time of the year and main reason of eutrophication is high organic load (Manna et. al, 2010).

9. Conclusion

Mangroves are the only woody halophytes dominated ecosystem situated at the confluence of land and sea, they occupy a harsh environment, being daily subject to tidal changes in temperature, water and salt-exposure and varying degree of anoxia (Alongi, 2008). The structure and function of the mangrove food web is unique, driven by both marine and terrestrial components. Little attention has been paid so far to the adaptive responses of mangrove biota to the various disturbances; however they are highly threatened even in this world heritage site, Sundarban. Recently eutrophication seems to be a global problem. The eutrophication of this tidal creek may have detrimental effect on the mangrove vegetation. Nutrient-enrichment i. e. eutrophication of the coastal zone increases the mortality of mangroves by enhancing shoot growth relative to root which makes them vulnerable to environmental stresses like salinity, draught that adversely affect plant water relationships (Catherine et al., 2009). Eutrophication as well as presence of toxic Dinoflagellates and Cyanophyceae in the tidal creek of Sundarban estuary definitely reveals the deteriorated status of the water quality.

Entire mangrove ecosystem of Sundarbans is fragile in nature due to many reasons such as erosion due to sea level rise, increase in salinity, pollution from non-point source like agricultural field wash and point sources like effluent from fishery, Kolkata sewage, loss of biodiversity due to continuous anthropogenic intervention, deforestation due to illegal felling and natural causes, eutrophication etc. Thus, a number of factors are operating in the Sundarbans. Apart from other aspects, different studies on Sundarbans gain its significance for throwing light on one of the most precious natural resource of this biogeographic region. This study indicates that ecosystem dynamics of Sundarbans may facilitate bioinvasion putting a question mark on the sustainability of mangroves. This work will create a roadmap to explore the ecosystem dynamics of Sundarban estuary, which is a great challenge in recent times.

10. Acknowledgement

We are grateful to Ministry of Earth Science, Government of India for the financial support to carry on this work. We express our sincere indebtness to Mr. Arijit Banerjee, Director, IESWM for his continuous support and interest in our work. We acknowledge the instrumental and infrastructural facility provided by UGC and DST, Government of India and ICZM project, West Bengal(World Bank)in the Department of Biochemistry, University of Calcutta. It would not be possible to carry out this immense work without the help and support of the local people of Sundarban. We express our inability to acknowledge them individually.

11. References

[1] Alongi DM. 2009. Life in Tidal Water. The Energetics of Mangrove Forests. Springer Science, pp 65-87.
[2] Alongi DM. 2002. Present state and future of the world's mangrove forests. Environmental Conservation. 29 (3): 331–349.
[3] Amann RI, Ludwig W, Schleifer KH. 1995. Phylogenetic identification and in situ detection of individual microbial cells without cultivation, Microbiol Rev, 59, 143.

Interplay of Physical, Chemical and Biological Components in Estuarine Ecosystem with Special
Reference to Sundarbans, India

207

[4] Anonymous, 2010 : Annual Report for the year 2009-10 of IESWM on the project entitled 'Oil Spill Trajectory Modeling along Sunderbans and Hooghly estuary' submitted to ICMAM Project Directorate, Ministry of Earth Sciences, Government of India

[5] APHA. 1998. Standard Methods for the Examination of Water and Wastewater. American Public Health Association, Washington, DC.

[6] Ariyadej, C., Tansakul, R., Tansakul, P. and Angsupanich, S. 2004. Phytoplankton diversity and its relationships to the physico-chemical environment in the Banglang Reservoir, Yala Province Songklanakarin J. Sci. Technol, 26(5) : 595-607.

[7] Atlas RM, Bartha R. 1993. Fundamentals and Applications. Benjamin_Cummings, Redwood City, Backwaters. Indian J. Mar. Sci., 3:46-50.

[8] Baidya AU & Chaudhury A. 1984. Ddtibution and abundance of zooplankton ina tidal creek of Sagar Islands, Sunderbanss, West Bengal. Envir. Ecol. 2: No. 4.

[9] Baidya, A. U. & Choudhury, A. (1984) Distribution and abundance of zooplankton in a tidal creek of Sagar Island, Sundarbans, West Bengal. Envir. & Ecol. 2: No. 4.

[10] Bakonyi T, Derakhshifar I, Grabensteiner L, Nowotny N: Development and evaluation of PCR assays for the detection of Paenibacillus larvae in honey samples: comparison with isolation and biochemical characterization. Appl. Environ Microbiol 2003, 69: 1504-1510.

[11] Bandyopadhyay, Ranjini and Sood, AK (2003) Effect of screening of intermicellar interactions on the linear and nonlinear rheology of a viscoelastic gel. In: Langmuir, 19 (8). pp. 3121-312

[12] Baron J, McKnight D, Denning AS. 1991. Sources of dissolved and particulate organic material in Loch Vale watershed, Rocky Mountain national park, Colorado, USA. Biogeochem 15:89-110.

[13] Bartilotti C, Macia A, Queiroga H. 2000. Larval fluxes at saco mangrove creek, Inhaca island (south Mozambique). In:Macrobenthos of Eastern African mangroves:life cycles and reproductive biology of exploited species. Final Report ERBIC 18-CT96-0127, Part B, Florence. pp 97-111.

[14] Bellinger EG. & Sigee DC. Freshwater algae: identification and use as bioindicator. 2010 John Wiley and Sons, -Science . pp 271 .

[15] Berounsky VM, Nixon SW. 1993 Rates of nitrification along an estuarine gradient in Narragansett Bay, Estuaries. 16: 718-730.

[16] Bhattacharyya, S., Pethick, J. & Sen Sarma K., 2011, Managerial response to sea level rise in the tidal estuaries of the Indian Sunderbans : a geomorphological approach, International Journal of Water policy– In press.

[17] Bianchi A, Giuliano L. 1995. Enumeration of Viable Bacteria in the Marine Pelagic Environment. Appl Enviro Microbiol. 62:174-177.

[18] Biddanda B, Opsahl S, Benner R. 1994. Plankton respiration and carbon flux through bacterioplankton on the Lousiana shelf, Limnol Oceanogr. 39: 1259-1275.

[19] Biswas H, Dey M, Ganguly D, De TK, Ghosh S, Jana TK: Comparative Analysis of Phytoplankton Composition and Abundance over a Two- Decade Period at the Land-Ocean Boundary of a Tropical Mangrove Ecosystem. Estuaries and Coasts 2010, 33:384-394.

[20] Blasco, F. 1977. Outlines of ecology, botany and forestry of the mangals of the Indian subcontinent. In:Chapman, V. (ed.) Wet Coastal Ecosystems. Ecosystems of the World. Vol. No. 1. Elsevier Scientific Publishing Co., Amsterdam. pp. 241-257.

[21] Boynton WR, Kemp WM, Keefe CW . 1982. A cornparative analysis of nutrients and other factors influencing estuarine phytoplankton production. Kennedy V(eds.). Estuarine comparisons. Academic Press, New York, pp. 69-90.

[22] Boynton WR, . Garber JH, Summers R and Kemp WM. 1995. Inputs, transformations, and transport of N and P in Chesapeake Bay and selected tributaries. Estuaries 18: 285-314.

[23] Boynton, WR, Kemp WM, Keefe CW. 1982. A cornparative analysis of nutrients and other factors influencing estuarine phytoplankton production. In: Kennedy, V, editor, Estuarine comparisons. Academic Press, New York, pp. 69-90.

[24] Cappuccino JG, Sherman N. 2007. Cultivation of Microorganisms. Microbiology: A Laboratory Manual. Cappuccino JG, Sherman N (eds.). Dorling Kindersley Publishers, India. ISBN 81-317-1437-3 129-134.

[25] Catherine EL, Marilyn CB, Katherine CM, Feller IC. 2009. Nutrient Enrichment Increases Mortality of Mangroves. PLoS ONE 2009, 4(5): e5600.

[26] Chisholm SW, Olson RJ, Zettle JER. Goericke, R., WaterburyJ. B. and Welshmeyer, N. A. (1988) A novel free-living prochlorophyte abundant in the oceanic euphotic zone. Nature. 334:340-343.

[27] Choudhury, A., and Chaudhury, A. B. 1994. Mangroves of the Sunderbanss, Volume one: India. IUCN-The World Conservation Union, Bangkok, Thailand.

[28] Cloern JE . 1996. Phytoplankton bloom dynamics in coastal ecosystems. A review with some general lessons from sustained investigation of San Francisco Bay. California. Rev Geophysics 33: 127-168.

[29] Cloern JE. 2001. Our evolving conceptual model of the coastal eutrophication problem. Mar Ecol Prog Ser. 210: 223–253.

[30] Cloern JE. 1996. Phytoplankton bloom dynamics in coastal ecosystems: A review with some general lessons from sustained investigation of San Francisco Bay. California. Rev Geophysics 33: 127-168.

[31] Cloern JL. 1999. The relative importance of light and nutrient limitation of phytoplankton growth: a simple index of coastal ecosystem sensitivity to nutrient enrichment. Aquatic Ecology 33: 3-16.

[32] Cloern JL. 1999. The relative importance of light and nutrient limitation of phytoplankton growth: a simple index of coastal ecosystem sensitivity to nutrient enrichment. Aquatic Ecology 33: 3-16.

[33] Cole JJ, Caraco NF. 1982. The pelagic microbial food web of oligotrophic lakes. In: Ford T (ed) Aquatic Microbiology. Blackwell Scientific, Cambridge, MA, pp 101–111.

[34] Conley DJ, Kaas F, Mohlenberg F, Rasmussen B, Windolf J (2000) Characteristics of Danish estuaries. Estuaries 23: 820-837.

[35] Conley DJ, Kaas F, Mohlenberg F, Rasmussen B, Windolf J . 2000. Characteristics of Danish estuaries. Estuaries 23: 820-837.

[36] Cowen JP. 2003. Fluids from aging ocean crust that support microbial life. Science. 299: 120-123.

[37] Crossland, C. J. D., Baird, D., Ducrotoy, J. P., Lindeboom, H., 2005. Thecoastalzone—a domain of globalinteractions. In:Crossland, C. J., Kremer, H. H., Lindeboom, H. J., Marshall Crossland, J. I., LeTissier, M. D. A. (Eds.), CoastalFluxesintheAnthropocene. InternationalGeosphere–BiosphereProgrammeSeries. Springer, pp. 1–37.

[38] Delft Hydraulics 1968. Sunderbans Delta Project Phase 1. Report to Rivers Research Institute, Gvt. West Bengal.

Interplay of Physical, Chemical and Biological Components in Estuarine Ecosystem with Special
Reference to Sundarbans, India

209

[39] Desai SR, Subash Chandran MD and Ramachandra TV. 2008. Phytoplankton Diversity in Sharavati River Basin, Central Western Ghats. The IUP Journal of Soil and Water. Sciences.

[40] Devassy, V. P. & Bhattathiri, P. M. A. 1974. Phytoplankton ecology of the Cochin Backwaters. Indian J. Mar. Sci., 3:46-50.

[41] Dur G, JS Hwang, S Souissi, LC Tseng, CH Wu, SH Hsiao, QC Chen. 2007. An overview of the influence of hydrodynamics on the spatial and temporal patterns of calanoid copepod communities around Taiwan. J. Plankt. Res. 29: 97-116.

[42] EEA: Europe's environment - The fourth assessment. State of the environment report No. 1/2007 2007, 452.

[43] Eyre BD . 1997. Water quality changes in an episodically flushed sub-tropical Australian estuary. A 50 year perspective. Mar chem 59: 177-187

[44] Falkowski PG, Barber RT, Smetacek V. 1998. Biogeochemical controls and feedbacks on ocean primary production. Science. 281: 200-206.

[45] Falkowski PG. 1980. hght-shade adaptation in marine phytoplankton. In: Falkowski, P. G, editor. Primary productivity in the sea. Environmental Science Research. Plenum Press, New York. pp. 99-119.

[46] Fenchel T. 1988. Marine plankton food chains. Ann. Rev. Ecol. Syst. 19: 19–38.

[47] Fieseler L, Quasier A, Schleper C, Hentschel U. 2006. Analysis of the first genome fragment from the marine sponge-associated, novel candidate phylum Poribacteria by environmental genomics. Environ Microbiol 2006, 8: 612-624.

[48] Finnish coastal waters. Monogr Boreal Environ Res. 31:1-57.

[49] Fogg GE. 1986. Picoplankton. Proc. R. Soc. Lond., 228: 1–30.

[50] Gle G, Amo BYD, Laborde SP, and Chardy P. 2008. Variability of nutrients and phytoplankton primary production in a shallow macrotidal coastal ecosystem (Arcachon Bay, France). Estuarine Coastal and Shelf Sciences, 76 : (), 642-656.

[51] Gopal B, Chauhan M. 2006 Biodiversity and its conservation in the Sunderbans Mangrove Ecosystem. Aquatic Sciences. 69:338-354.

[52] Gopinathan, CP. 1972. Seasonal abundance of phytoplankton in the Cochin Backwaters. J. Mar. biol. Ass. India. 14: 568-577.

[53] Gouda, R. & Panigraphy, R. C. 1996. Seasonal distribution of phytoplankton in the surf waters off Gopalpur, east coast of India. Indian J. Mar. Sci., 25:146-150.

[54] Graneli E, Wallstrom K, Larsson U, Graneli W, Elmgren R. 1990. Nutrient limitation of primary production in marine ecosystems in the Baltic Sea. Ambio. 19: 142-151.

[55] Gupta GVM, Thottathil SD, Balachandran, KK, Madhu NV, Madeswaran P, Nair S. 2009. CO2 Supersaturation and net heterotrophy in a tropical estuary (Cochin, India) : Infiuence of Antropqgenic effect. Ecosystems. 12(12): 1145-1157.

[56] Harding LW Jr. 1993. Long-term trends in the distribution of phytoplankton in Chesapeake Bay: rolcs of light, nutrients and streamflow. Mar Ecol Prog Ser. 104: 267-291.

[57] Harding LW, Perry ES Jr. 1997. Longterm increase of phytoplankton biomass in Chesapeake Bay, 1950-1994. Mar Ecol Prog Ser. 157:39-52.

[58] Harris GP. 1978. Photosynthesis, productivity and growth: The physiological ecology of phytoplankton. Arch. Hydrobiol. Beih. Ergeb. limnol. 10:. 10: 1-71.

[59] HELCOM: Environment of the Baltic Sea area 1994-1998. Balt Sea Environ Proc No. 82B 2002, 214.

[60] High Diversity. Trophial conservation Biology, . Blackwell Publishing Ltd, 1-32, ISBN 978-1-4051-5073-6.

[61] Hobbie JE, Daleyr RJ, Jasper S. 1976. Use of Nuclepore Filters for Counting Bacteria by Fluorescence Microscopy. Appl Environ Microbiol., 33:1225-1228.

[62] Hobbie JE, Kling GW, Rublee PA. 1996. Controls of bacterial numbers and productivity in an arctic lake. Bull Ecol Soc Am 77:197.

[63] Holben WE, Tiedje JM. 1988. Applications of nucleic acid hybridization in microbial ecology, Ecology 1988, 69, 561.

[64] Holguin G, Vazquez P, Bashan Y. 2001. The role of sediment microorganisms in the productivity, conservation, and rehabilitation of mangrove ecosystems: an overview. Biol Fertil Soils . 33:265–278.

[65] Hopkmson CS, Sherr B, Wiebe W. 1989. Size fractionated metabolism of coastal microbial plankton, Mar Eco Prog Ser. 51: 155-166.

[66] Hughes TP, Baird AH, Bellwood DR, Card M, Connolly SR. 2003. Climate change, human impacts and the resilience of coral reefs. Science. 301: 929–933.

[67] Hwang JS, HU Dahms, V Aleeksev. 2008. Novel nursery habitat of hydrothermal vent crabs. Crustaceana 81: 375-380.

[68] International Newsletter of Coastal Management-Intercoast Network, Special J. Mar. biol. Ass. India. 14: 568-577.

[69] Jansson M, Bergstrom AK, Blomqvist P, Drakare S. 2000. Allochthonous organic carbon and phytoplankton/bacterioplankton production relationships in lakes. Ecology 81:3250– 3255.

[70] Jansson M, Blomqvist P, Jonsson A, Bergstrom AK. 1996. Nutrient limitation of bacterioplankton, autotrophic and mixotrophic phytoplankton, and heterotrophic nanoflagellates in Lake Ortrasket, a large humic lake in northern Sweden. Limnol Oceanogr 41:1552–1559.

[71] Jara-Marini ME, Soto-Jimenez MF, Paez-Osuna F. 2009. Trophic relationships and transference of cadmium, copper, lead and zinc in a subtropical coastal lagoon food web from SE Gulf of California. Chemosphere. 77(10): 1366-1373.

[72] JGFOS Protocol. 1994. Protocol for the Joint Global Flux Study (JGFOS) Core Management. 178.

[73] John WD, Charles AS, Michael W K, Alejandro YA. 1989. "Estuarine Ecology. " Wiley-Interscience; 1 edition. ISBN 0-10-0471062634.

[74] Johnson PW and Sieburt JM. 1982. In situ morphology and occurrence of eukariotic phototrophs of bacterial size in the picoplankton of estuarine and oceani waters. J. Phycol. 8:318–327.

[75] Jones RA, Fred Lee G. 1982. Recent Advances in Assessing Impact of Phosphorus load on eutrophication-related water quality. Water Research 1982, 16:503-515.

[76] Jones RI. 1992. The influence of humic substances on lacustrine planktonic food chains. Hydrobiol 229:73–91.

[77] Jones RI. 1998. Phytoplankton, primary production and nutrient cycling. In: Hessen DO, Tranvik LJ (eds) Aquatic Humic Substances — Ecology and Biogeochemistry. SpringerVerlag, Berlin, pp 145–175.

[78] Kannan, L. & K. Vasantha, 1992. Microphytoplankton of the Pitchavaram mangals, southeast coast of India: Species composition and population density. In Jaccarini,

Interplay of Physical, Chemical and Biological Components in Estuarine Ecosystem with Special
Reference to Sundarbans, India

211

V. & E. Martens (eds), The Ecology of Mangrove and Related Ecosystems.
Hydrobiologia

[79] Karner MB, Delong EF, Karl DM. 2001. Archaeal dominance in the mesopelagic zone of
the Pacific Ocean. Nature. 409: 507-510.

[80] Kauppila P. 2007. Phytoplankton quantity as an indicator of eutrophication in Finnish
coastal waters. Monogr Boreal Environ Res. 31:1-57.

[81] Kausch H. 1990. Biologtcal processes m the estuarine environment In Estuarine Water
Quality Management edited by Mihaehs W Spnnger-Verlag, Berhn, Coastal and
Eatuarine Studies. 36: 353-361.

[82] Kemp WM and. Boynton WR. 1992. Benthic-pelagic interactions: Nutrient and oxygen
dynamics, p. 149-209. In D. E. Smith, M. Leffler, and G. Mackiernan (eds.), Oxygen
Dynamics in the Chesapeake Bay: A Synthesis of Recent Research. Maryland Sea
Grant, College Park, Maryland.

[83] Kemp WM, Bonyton WR, Adolf JE, Boesch DF, Boicourt WC, Brush G, Cornwell JC,
Fisher TR, Glibert PA, Hugy JD, Harding LW, Houde Jr. ED, Kimmel DG, Millar
WD, Newell RIE, Roman MR, Smith EM. 2005. Stevenson JC: Eutrophication of
Chesapeake Bay: Historical trends and ecological interactions. Mar eco Pro Ser. 303:
1-29.

[84] Kemp WM, Sampou PA, Garber J, Tuttle J, Boynton WR. 1992. Seasonal depletion of
oxygen from bottom waters of Chesapeake Bay roles of benthic and planktonic
respiration and physical exchange processes. Mar Eco Prog Ser. 85: 137-152.

[85] Kiirikki M, Lehtoranta J, Inkala A, Pitkänen H, Hietanen S, Hall POJ, Tengberg A,
Koponen J, Sarkkula J. 2006. A simple sediment process description suitable for 3D-
ecosystem modelling - Development and testing in the Gulf of Finland. J Mar Syst.
61:55-66.

[86] Knudson M. 1901. Hydrographical tables. G. E. C. Gad Copenhagen 63.

[87] Kroer N. 1993. Bacterial growth efficiency on natural dissolved organic matter. Limnol
Oceanogr 38:1282–1290.

[88] Lahiri, R. 1973. Management Plan of Tiger Reserve in Sunderbanss, West Bengal, India.
Department of Forests, West Bengal. 101 pp.

[89] Lapointe BE. 1997. Nutrient thresholds for bottom-up control of macroalgal blooms on
coral reefs in Jamaica and southeast Florida. Limnol Oceanogr. 42: 1119–1131.

[90] Lee RE. 1999. Psychology. Cambridge University Press, New York, 614 pp.

[91] Leila J. Hamdan, and Robert B. Jonas 2007. Seasonal and interannual dynamics of free-
living bacterioplankton and microbially labile organic carbon along the salinity
gradient of the Potomac River. Estuaries and Coasts. 29(1):40-53.

[92] Levine J, Brewer S, Bertness M. 1998. Nutrients, competition and plant zonation in a
New England salt marsh. J Ecol 86: 285–292.

[93] Mani, P., Krishnamurthy, K. and Palaniappan, R. 1986. Ecology of phytoplankton
blooms in the Vellar Estuary, east coast of India. : Indian J. Mar. Sci. 15(1): 24-28.

[94] Manna S, Chaudhuri K, Bhattacharyya S, Bhattacharyya M. 2010. Dynamics of
Sunderbans estuarine ecosystem: eutrophication induced threat to mangroves.
Saline Systems. 6:8

[95] Margalef DR. 1951. Diversidad de especies en les communideades natural Public
Institutte of Biologic Barcelonia . 1951; 9: 5 – 27.

[96] McLusky, D. S., Elliott, M., 2004. The Estuarine Ecosystem: Ecology, Threats and Management, third ed. Oxford University Press, Oxford, pp 216.

[97] Mishra, S., and Panigrahy, RC, 1995. Occurance of diatomblooms in Bahuda estuary, ast coast of India. Indian J. Mar. Sci., 24: 99-101.

[98] Morgan JAW, Winstanley C: Microbial biomarkers. In: van Elsas, JD, Trevors JT, Wellington EMH (Eds.), Modern Soil Microbiology. Marcel Dekker, New York 1997, 331-352.

[99] Nixon SW. 1995. Coastal marine eutrophication: a definition. social causes and future concerns. Ophelia. 31: 199-219.

[100] Nwankwo DI, Akinsoji A. 1992. Epiphyte community on water hyacinth Eichhornia crassipes (Mart.). Solms. in coastal waters of southwestern Nigeria. Arch. Hydrobiol. 124(4): 501-511.

[101] Oachs CA, Cole JL, Likens GE. 1995. Population dynamics of bacterioplankton in an oligotrophic lake. J Plank Res 17:265-391.

[102] Owens. NJP: 1986. Estuarine nitrification: a naturally occurring fluidized bed reaction". Est Coast Shelf Sci. 22: 31-44.

[103] Pace NR, Stahl DA, Lane DJ, and Olsen GJ: The analysis of natural microbial populations by ribosomal RNA sequences, Adv Microb Ecol 1986, 9, 1.

[104] Paerl HW. 1997. Coastal eutrophication and harmful algal blooms: Importance of atmospheric deposition and groundwater as "new" nitrogen and other nutrient sources. Limnol Oceanogr. 42: 1154-11.

[105] Pakulsht JD, Benner R, Amon R, Eadie B, Whitledge T. 1995. Community metabolism and nutrient cycling in the Mississippi River plume: evidence for intense nitrification at intermediate salinities, Mar Eco Prog Ser. 117: 207-218.

[106] Palmer M C. 1969. A Composite Rating of Algae Toleratng Organic Pollution. Journal of Phycology 5: 78-82.

[107] Pargiter, Frederick Eden, 1934 : A revenue history of the Sunderbanss, Volume-I (From 1765 to 1870) pp1-415, republished by Government of West Bengal in 2002.

[108] Patrick M, Ysebaert T, Damme SV, Bergh EV, Maris T & Struyf E. 2005. The Scheldt estuary: a description of a changing ecosystem Hydrobiologia. 540:1-11

[109] Paula, J. 1998. larval retention and dynamics of the prawns Palaemon longirostris H. Milne Edwards and Crangon crangon Linnaeus (Decapoda, Caridea) in the Mira estuary, Portugal, Inv. Reprod. Devel. 33:221-228

[110] Pennock JR. 1985. Chlorophyll distributions in the Delaware Estuary: Regulation by light-limitation. Estuar. st. Shelf Sci 21: 711-725.

[111] Perry R. 2003. A Guide to Marineplankton of southern California. Mar Sci Center, Ocean Globe 3rd edition, 1-17.

[112] Phytoplankton of Bhagirathi-Hooghly estuary: An illustrative account. Ind. Biol.,

[113] Pillay TVR. 1958. Biology of the Hilsa sp. Ind J Fish. 5:201-207.

[114] Pomeroy LR, Johannes RE. 1968. Occurrence and respiration of ultraplankton in the upper 500 meters of the ocean. Deep-Sea Res. 15: 381-391.

[115] Porter KG, Feig YS. 1980. The use of DAPI for identifying and counting aquatic microflora. Limnol Oceanogr. 25: 943-948.

[116] Prabhahar. C, Saleshrani. K and Enbarasan R. 2011. Studies on the ecology and distribution of phytoplankton biomass in Kadalur coastal zone Tamil nadu, India. Curr. Bot. 2(3):26-30, 2011.

Interplay of Physical, Chemical and Biological Components in Estuarine Ecosystem with Special
Reference to Sundarbans, India

213

[117] Rae R, Vincent WF. 1998. Phytoplankton production in subarctic lake and river ecosystems: development of a photosynthesis– temperature–irradiance model. J Plank Res 20:1293–1312.

[118] Ram ASP, Nair S and Chandramohan D. 2003. Seasonal Shift in Net Ecosystem Production in a Tropical Estuary. Limnol. & Oceanogra. 48(4): 1601-1607.

[119] Ray S. 2008. Comparative study of virgin and reclaimed islands of Sunderbans mangrove ecosystem through network analysis. Eecol Model. 215:207-216.

[120] Reservoir, Yala Province Songklanakarin J. Sci. Technol, 26(5) : 595-607.

[121] Reynolds CS. 1984. Phytoplankton periodicity: The interactions of form, function and environmental variability. Freshwater Biology. 14: 111–142.

[122] Reynolds, C. S. 1993. Scales of distribution and their role in plankton ecology, Hydrobiol., 249, 157 – 171.

[123] Richardson K. 1997. Harmful or exceptional phytoplankton blooms in the marine ecosystem. Advances in Marine Biology 31: 301– 385.

[124] Riley GA. 1967. The plankton of estuaries. In: Lauff, G. A. (ed.) Estuaries. AAAS Publ., Washington D. C., p. 316-328.

[125] Riley GA. 1967. The plankton of estuaries. In: Lauff, G. A. (ed.) Estuaries. AAAS Publ., Washington D. C., p. 316-328.

[126] Rissik D: Plankton: A Guide to their Ecology and Monitoring for Water Quality. Suthers IM, Rissik D (eds.). CSIRO Publishing, Collingwood, Australia; 2009. ISBN: 9780643090583.

[127] Robertson, A. I. and S. J. M. Blabber (1992). plankton, Epibenthos and fish communities. In: Tropical Mangrove Ecosystems (A. I. Robertson and D. M. Alongi Eds.). Coastal Estuar. Stud., 41: 173-224.

[128] Rosenberg R, Elmgren R, Fleischer S, Johnsson P, Persson G, Dahlin H. 1990. Marine eutrophication studies in Sweden. Ambio. 19:102-108.

[129] Ryther JH, 1969. Photosynthesis and fish production in the sea. Science 166:72-76. 969). Photosynthesis and fish production in the sea. Science. 166:72-76.

[130] Ryther JH, 1969. Photosynthesis and fish production in the sea. Science 166:72-76. 969). Photosynthesis and fish production in the sea. Science. 166:72-76.

[131] S. K. Mukhopadhyay, H. Biswas, T. K. De and T. K. Jana, 2006, Fluxes of nutrients from the tropical River Hooghly at the land-ocean boundary of Sunderbanss, NE Coast of Bay of Bengal, India, Journal of Marine System 62: 9 - 21.

[132] Santelli CM, Orcutt BN, Banning E, Bach W, Moyer CL, Sogin ML, Staudigel H, Edwards KJ: Abundance and diversity of microbial life in ocean crust. Nature 2008, 453: 653-657 doi: 10. 1038.

[133] Santoro, A. E. ;Francis, C. A. ;de Sieyes, N. R. ;Boehm, A. B. 2008. Shifts in the relative abundance of ammonia-oxidizing bacteria and archaea across physicochemical gradients in a subterranean estuary. Environmental Microbiology. 10(4): 1068-1079.

[134] Santra S C and Pal U C 1989 Phytoplankton of Bhagirathi- Hooghly estuary. An illustrative account. Indian Biology. 29: 1-27.

[135] Santra, S. C., Das, U. C., Sima Sen, T. M., Rita Saha, Dutta, S. & Gosh Dastidae, P. 1989. abundance of ammonia-oxidizing bacteria and archaea across

physicochemical gradients in a subterranean estuary. Environmental Microbiology. 10(4): 1068-1079.

[136] Sanyal, P. and Bal, A. R. 1986. Some observations on abnormal adaptations of mangrove in Indian Sunderbanss. Indian Soc. Coastal agric. Res. 4: 9-15.

[137] Sarkar A., S. Sengupta S, McArthur JM, Ravenscroft P, M. K. Bera, Ravi Bhushan, A. Samanta and S. Agrawal; 2009. Evolution of Ganges–Brahmaputra western delta plain: Clues from sedimentology and carbon isotopes. Quaternary Science Reviews., 28: 2564 - 2581.

[138] Scheffer M, Carpenter S, Foley JA, Folke C, Walker B. 2001. Catastrophic shifts in ecosystems. Nature. 413: 591–596.

[139] Sieburth, JMcN., SmatacekV, . and Lenz, J. 1978. Pelagic ecosystem structure: heterotrophic compartments of the plankton and their relationship to plankton size fractions. Limnol. Oceanogr., 23, 1256–1263.

[140] Sodhi NS, Brok BW, Bradshaw CJA. 2007. Diminishing Habitats in Regions of High Diversity. Trophial conservation Biology, . Blackwell Publishing Ltd, 1-32, ISBN 978-1-4051-5073-6.

[141] Sogin ML, Morrison HG, Huber JA, Welch DM, Huse SM, Neal PR, Arrieta JM, Herndl GJ. 2006. Microbial diversity in the deep sea and the underexplored "rare biosphere".

[142] Spalding, M. 1997. The global distribution and status of mangrove ecosystems. International Newsletter of Coastal Management-Intercoast Network, Special edition, 1 : 20-21.

[143] Spiers, A. G., 1999. Review of international/continental wetland resources. In: C. M. Finlayson and A. G. Spiers (eds.), Global Review of Wetland Resources and Priorities for Wetland Inventory. Wetlands Internation Publication 13. Wetlands International, Springer Science, pp 65-87.

[144] Stanley, D. J. ; Hait, A. K. ; 2000 Holocene depositional patterns, neotectonics and Sunderbans mangroves in the western Ganges-Brahmaputra delta. Journal of Coastal Research, 2000 – pp26-39

[145] Steele HL, Streit WR: Metagenomics. 2005. advances in ecology and biotechnology. FEMS Microbiol Lett 2005, 247: 105-111.

[146] Stockner, JG. and Antia, NJ. 1986. Algal picoplankton from marine and freshwater ecosystems: A multidisciplinary perspective. Can. J. Fish. Aquat. Sci. 43: 2472–2503.

[147] Stolte W, Mccollin T, Noordeloos A. and. Riegman R. 1994. Effect of nitrogen-source on the size distribution within marinephytoplankton population. Journal of Experimental Marine Biology and Ecology . 184: 83–97.

[148] Struyf E. 2005. The Scheldt estuary: a description of a changing ecosystem Hydrobiologia 540:1–11

[149] Sundh I, Bell RT. 1992. Extracellular dissolved organic carbon released from phytoplankton as a source of carbon for heterotrophic bacteria in lakes of different humic content. Hydrobiol 229:93–106.

[150] Teague, C. C. ; Barrick, D. E. ; Lilleboe, P. M. ; Cheng, R. T. ; Geoscience and Remote Sensing Symposium, 2007, IGARSS 2007, IEEE International 23-28 July 2007 p 2491 – 2494, Barcelona

Interplay of Physical, Chemical and Biological Components in Estuarine Ecosystem with Special Reference to Sundarbans, India

215

[151] Tomas CR. 1997. Identifying marine phytoplankton. Academic Press, Harcourt Brace and Company, Torpnt. 858

[152] Tomlinson, B. P. 1986. The botany of mangrove. Cambridge University Press. London. UK.

[153] Tranvik LJ. 1988. Availability of dissolved organic carbon for planktonic bacteria in oligotrophic lakes of different humic content. Microb Ecol 16:311–322.

[154] Tringe SG, Rubin EM; Metagenomics. 2005. DNA sequencing of environmental samples. Nat Rev Gene. 6: 805- 814.

[155] Trivedi RK and Goel PK. 1980. Chemical and Biological methods for water pollution studies. Environmental Publication (India), pp. 250.

[156] Tseng LC, R Kumar, HU Dahms, QC Chen, JS Hwang. 2008a. Monsoon driven succession of copepod assemblages in coasta l waters of northeastern Taiwan strait. Zool. Stud. 47: 46-60

[157] Tulonen T, Kankaala P, Ojala A, Arvola L (1994) Factors controlling production of phytoplankton and bacteria under ice in a humic, boreal lake. J Plank Res 16:1411–1432.

[158] Turner RE, Rabalais NN (2003) Linking landscape and water quality in the Mississippi River basin for 200 years. Bioscience 53: 563-572.

[159] Turner RE, Rabalais NN. 1991. Changes in Mississippi River water quality this century, Bio Sci. 41:140-147.

[160] UNESCO. 1978. Monograph and Oceanographic methodology. Phytoplankton Manual. Sournia, A (eds.). In UNESCO Technicalpaper No. 28, 337.

[161] Valdes-Weaver, LM., Pichler MF. Pinckney J T., Howe KE. Rossignol K and. Paerl HW. 2006. Long-term temporal and spatial trends in phytoplankton biomass and class-level taxonomic composition in the hydrologically variable Neuse-Pamlico estuarine continuum, North Carolina U. S. A. Limnology and Oceanography. 51(3): 1410–1420.

[162] Valiela I, McClelland J, Hauxwell J, Behr PJ, Hersh D, Foreman K. 1997. Macroalgal blooms in shallow estuaries: Controls and ecophysiological and ecosystem consequences. Limnol Oceanogr. 42:1105-1118.

[163] Ventrick EL, Hayward TL. 1984. Determining chlorophyll on the 1984 CalCOFI surveys. CalCOFI Rep. 25:74-79.

[164] Volunteer Estuary Monitoring: A Methods Manual. March 2006. EPA-842-B-06-003.

[165] Waniek, J. J. and N. P. Holliday: 2006 Large – scale physical control on phytoplankton growth in the Irminger Sea, Part II: Model study of the physical and meteorological

[166] Ward DM, Bateson MM, Weller R, Ruff-Roberts AL. 1992. Ribosomal RNA analysis of microorganisms as they occur in nature, Adv Microb Ecol. 12, 219.

[167] Weiss R. 1993. Dynamics of autotrophic picoplankton in marine and freshwater ecosystems. Adv. Microb. Ecol. 13: 327.

[168] Whitman WB, Coleman DC, Wiebe WJ. 1998. Prokaryotes: the unseen majority. Proc Natl Acad Sci USA. 95:6578-6583.

[169] Williams RB. 1972. Annual phytoplankton production in a system of shallow temperate estuaries. In: Chabreck, RH. (ed.) Proceedmgs of the Coastal Marsh and Estuary Management Synlposium. Louisiana State Univ. Press, Baton Rouge, p. 59-89.

[170] Willlams PJ. 1981. Microbial contribution to overall marine plankton metabolism: direct measurements of respiration. Ocranol Acta. 4: 359-364.

[171] Wofsy SC. 1983. A simple model to predict extinction coefficients and phytoplankton biomass in eutrophic waters. Limnol, Oceanogr. 28: 1144-1155.

10

Diatoms as Indicators of Water Quality and Ecological Status: Sampling, Analysis and Some Ecological Remarks

Gonzalo Martín and María de los Reyes Fernández
University of Seville
Spain

1. Introduction

Diatoms are amazing microscopical algae whose typical feature is a siliceous coverage, called frustule, extremely diverse in shape. Diatoms live in almost all types of superficial waters. Depending on their habitats, diatoms are either planktonic (living suspended on the water), benthic (growing associated to a substrate), or both planktonic and benthic.

On one hand, planktonic diatoms usually have fine frustules and/or long appendices to facilitate floatability. Many marine diatoms, like the genus *Bacteriastrum* Shadbolt, are good examples of this characteristic. The chain formation facilitates the planktonic life too, such as in *Skeletonema costatum* (Greville) Cleve or *Fragilaria crotonensis* Kitton.

On the other hand, benthic diatoms do not need delicate structures because they live on a substrate. Therefore they do not have to worry about sinking. They can be motile when living on sediment like many species of *Navicula* Bory, grow closely attached to a substrate like *Cocconeis* Ehrenberg (Fig. 1), or live on top of auto-produced mucilaginous stalks as many species of *Gomphonema* Ehrenberg do.

The algal species that develop in an area depend on different environmental factors: salinity, temperature, pH, water velocity, shading, depth, availability of substrata to grow on, water chemistry, etc. Thus, the species that can be found in a water body will inform about some characteristics of the water. Because algae are good indicators of the features of the water, they are used to monitor water quality. Among algae, diatoms have the advantage of being easily identifiable at the species level without the need of cultures and they are very easy to collect and store due to their hard frustule. The ecological requirements of many diatom species are known and therefore many diatom-based indexes of water quality have been developed.

Such indexes have been created for benthic diatoms of continental running waters. There is a reason why benthic diatoms are more reliable for this purpose—they remain in their location unless disturbed by some catastrophic event, like a big flood. This means that they reflect the characteristics of the water from the area in which they live.

On the contrary, planktonic diatoms "get old with the water"—as the water runs, they move with it. Therefore, when a researcher collects planktonic diatoms that grew upstream, such algae may reflect other water characteristics than the ones in the sampling site.

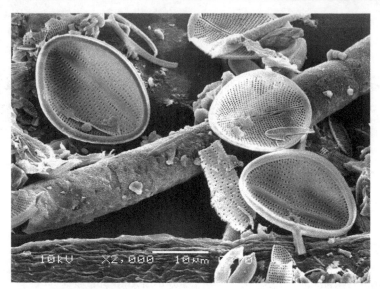

Fig. 1. Four valves of *Cocconeis* Ehrenberg.

In this chapter we will explore the use of diatoms to monitor water quality. To achieve this, this chapter has two main objectives:

1. To be a guide for those who work with this type of algae: The chapter shows how sampling must be carried out and how samples have to be treated and analysed.
2. To warn about the risk of using the indexes alone, without any further information. There are some situations that can lead to incorrect conclusions about the health of an ecosystem if only the values of the indexes are observed. For this reason, we show two cases in which diatom indexes reflect something different from the actual state of the ecosystem.

2. Water quality and ecological status of a water body

In recent decades, a significant effort has been put forth all over the world to assess water quality attending to not only to chemical parameters (nutrients, metals, pesticides, etc.), which are obviously important, but also to biological indicators. In fact, one of the undesirable consequences of pollutants is their effect on biota. In this case, the direct study of the effects of pollution on biota is of great interest.

However, it is not so easy to define the extent to which an ecosystem is damaged. There are obvious situations in which heavily polluted waters give the ecosystems such a structure and characteristics that there is no doubt about the heaviness of the human pressures and no effort is necessary to get conclusions about the health of the system. However, there are intermediate situations where it is difficult to draw a conclusion about the water quality without further evaluation.

In this context, the European Water Framework Directive (2000) has a simple philosophy, easy to understand but difficult to put in practice: An ecosystem should be evaluated according to how it was expected to be under no human pressure. Such an ecosystem would be considered to have a good ecological status and could be used as a pattern – the so-called reference conditions – with which other similar ecosystems can be compared. There are five levels of quality, which are identified by colours, according to the closeness of the ecosystem to its reference condition: red, very poor; orange, poor; yellow, moderate; green, good; blue, excellent. Despite these five standards of water quality levels, it is hard to know how each ecosystem would be under pristine conditions, which leaves much work to be done.

There are different biological indicators that can be used to determine the ecological status of a water body: macroinvertebrates, aquatic plants, and algae, especially diatoms, are all used in water quality research (Triest et al., 2001; Hering et al., 2006; Torrisi et al. 2010). The ecological information given by diatoms is usually summed up through one or more diatom-based indexes, which indicate the trophic level with just one number. However, the indexes alone do not determine the water quality. Moreover, the water quality is not the only factor that determines the ecological status. Other factors such as the structure of the vegetation or the hydrodynamics of the ecosystem must also be taken into account.

In summary, the use of the values of the diatom indexes alone may not be enough to determine if an ecosystem has got a good status or not and sometimes can be insufficient even to determine the water quality, as will be shown at the end of this chapter.

3. Diatoms sampling

Currently, the diatom sampling process is mostly normalised, since there are norms to ensure the accuracy of the procedure. The European Standard EN 13946 (2003), as well as the recomendations provided by Kelly et al. (1998) are well known and widely followed.

In most cases, the diatom sampling is carried out in rivers in order to monitor water quality through the calculation of one or more indexes based on these algae. Firstly, we will focus on rivers, and secondly, we will show some other variations to sample diatoms in wetlands. The following sections summarize the sampling procedure in order to serve as a quick guide.

3.1 Sampling rivers

Sampling sites must be located up and downstream of any impact in the basin of which the researchers are aware but the final selection of the sampling sites depends on the objectives of the work.

The sampling must be carried out in a lightly shadowed by riparian vegetation segment of the river of 10 m length or more. This segment has to include rapids, since stretches of river with very slow currents (<20 cm/s) allow the buildup of loosely attached diatoms, silt, and other debris. Samples have to be collected in rapids at a distance from the riverbanks to avoid waters somehow isolated from the main current that would not reflect the characteristics of the site along with any substrata that may had recently remained out of the water and could contain aerophilous diatoms.

The next step is to choose the type of substrate to sample. There are many works explaining how the sampled substrata can influence the species that can be found on the rivers

(Danilov & Ekelund, 2001; Albay & Akcaalan, 2003; Potapova & Charles, 2005; Towsend & Gell, 2005). For example, some diatoms, such as *Epithemia adnata* (Kützing) Brebisson or *Cocconeis pediculus* Ehrenberg are more likely to be found growing on plants or macroalgae rather than on rocks.

Natural rocks are the best substrate for sampling springs (Round, 1991). However, researchers should avoid any large rock difficult to remove from the river because it would make the sampling process too difficult. Pebbles are not appropriate either, because they might have come recently from upstream and they can mislead the sample. The best choice is an average-size rock of about 15 to 30 cm. However, when this type of substrata is not available, the researcher should take samples in the following order of preference: natural rocks, artificial surfaces (such as bridge pillars), artificial substrates, and plants.

Artificial substrates must be placed in the riverbed at least one month before the sampling is executed to ensure the proper development of a mature algal community. Some researches use slides, which have the advantage of giving a known surface and let the direct observation of the algal growing and the biofilm architecture under a microscope. However, it is preferable to use tiles or bricks in routine biomonitoring work because they provide rough surfaces resembling natural rocks that facilitate the settlement of algae. Unfortunately, all these substrates are often lost in the riverbed due to strong currents and human interaction with the river. Moreover, these artificial substrates can be found completely covered of fine sediments instead of algae, which is quite common since these substrata are likely to be left in rivers whose riverbed has only fine sediment and no natural rocks. A stake can be driven into the sediment through the holes of several bricks, which should be piled up at the bottom of the river. It ensures a vertical surface colonizable by algae on which little mud would sediment. Moreover, it will be easy to localize when returning to the river.

When the researcher decides to sample natural rocks, he or she should randomly collect between five and ten of them. Then, the surface of the rocks should be scrapped off and removed with a toothbrush. The brushed area of the stones, as well as the thoothbrush covered with algae, can be cleaned on a plastic tray with water from the river if it is clear or with tap water if the river runs muddy. The obtained brownish suspension constitutes the sample. Tiles and bricks must be treated as natural rocks.

When sampling diatoms growing on plants, the researcher can use either emergent plants– also called helophytes, such as *Thypha* L. or *Phragmites* Adans. – or submerged plants – the so-called hygrophyes, like *Ceratophyllum* L. or *Myriophyllum* L. Not all plant species facilitate the growing of the same diatoms, in quantity and diversity. Some plants are harder to be colonished by algae for many reasons, such as being too smooth for the algae to grow, etc. For example, *Typha* L. typically exhibits a thicker biofilm than *Scirpus* L. In general, it is recommended to always sample on the same plant species in order to make the data comparison from different sites easier. It is often impossible to do this, since the elected species may not grow in all sampling sites, but this is not a matter of extreme importance, according to Cejudo-Figueiras et al. (2010)

The algae growing on plants have to be collected by gathering leaves that are alive and stem sections, discarding either those parts that had been recently out of the water or those that are near the bottom and covered by sediment. Cuttings of about 5-cm length must be carefully placed together in a bottle filled with tap water. The stems and leaves have to be gently scraped with a cover slip in the laboratory, while softly washed with water. The obtained brown or greenish suspension contains the diatoms that will be used in the research. This procedure is suitable with helophytes – *Typha*, *Scirpus*, *Phragmites*, etc. – but not with fine and delicate hygrophytes such as *Myriophyllum*, which would be completely smashed with the coverslip. In that case, some fragments of plants must be introduced into a plastic recipient with some water (from the river if it is clear or tap water otherwise). The recipient must be closed and shaked vigorously. The water will become muddy, brownish or greenish. The plants can be removed out of the jar and replaced with other fragments. After repeating this operation three or four times, the water from the recipient will constitute the sample and it will be blurred because of all the epiphitic algae.

3.2 Sampling wetlands

Small springs are quite easy to sample, since their riverbed is not muddy and usually have many rocks of different sizes that are easily removable. However, wetlands, as well as wide rivers, only have plants and sediments as the only available substrata, and in some cases even plants are absent. If there are plants, they should be treated as described above for rivers. As a general rule, plants have demonstrated to be better than stones in wetlands because plants provide a vertical surface mostly free of mud (Blanco et al., 2004). Stones, if present, typically have a thick layer of sediment, making them useless for sampling. However, the allocation of artificial substrata is a good choice too, if they are disposed vertically (Sekar et al., 2004).

Sediment is not a suitable substrate because diatoms growing among its particles can get nutrients from the intersticial water, having an extra source of nutrients that other algae living out of the sediment lack. This means that trophic levels would be overestimated when analysing diatoms from sediments. That is why rocks are preferred, since they cannot provide extra nutrients to the algae, forcing them to obtain the nutrients needed from the open water.

Nevertheless, some studies may recommend sampling sediment if the goal is not the application of water quality indexes but the knowledge of the algal flora. For example, there are wide marshlands in which there are large extensions of sediment (under the water or just in wet areas) free of plants. Sometimes the primary production of this sediment is very important for the whole system and, very often, a greenish highly productive biofilm can grow on it. If the researcher wants to study the algal composition of this marshland, attention must be focused on sediments, which is what characterizes these types of ecosystems, rather than any other type of substrata.

What is the best way to sample algae from sediments? If researchers want to perform quantitative studies they must sample a known surface. Half of a Petri dish must be driven into sediment to take the superficial layer, using the aid of a scrapper to facilitate the process. After homogenising the sample at the laboratory, small subsamples can be taken from it to count algae under the microscope or to measure chlorophyll values. Any subsample must be extrapolated to the volume or weight of the whole sample and to the area of the Petri dish.

Analysing sediment under the microscope is terribly weary and exasperating. If the objective of the study is to identify the species with a rough estimate of their proportions, there is a strategy that not only facilitates this task but also ensures that only living diatoms are observed. Diatoms living in sediment, like many species of *Nitzschia* Hassall, *Gyrosigma* Hassall or *Navicula* Bory (Fig. 2) are motile, which implies that the sample – or a subsample after homogenising the whole sample – can be left in the Petri dish with a thin layer of water for about 12 hours under good illumination, which acts like a lure attracting diatoms towards the light. A thin green biofilm will appear on the mud that can be carefully picked up and disposed on a slide with a drop of water to be observed under the microscope. Moving algae may adhere to some cover slips if they are left on the wet sediment, although they tend to accumulate around the slips better than under them. Therefore, the analysis of these cover slips is easier because they will have less mud.

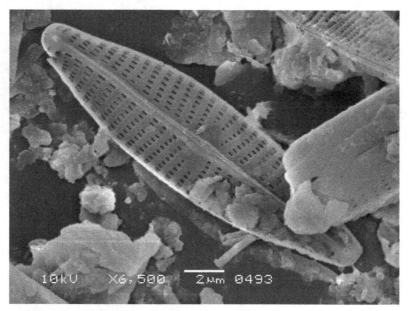

Fig. 2. A single valve of *Navicula* among particles of sediment.

In this case previously mentioned, diatoms have to be alive in order to move among the particles of the sediment. In any other case, if the collected material is not going to be treated in the lab within one day– no matter the substrate or the type of sampled ecosystem— it must be homogenized and fixed with 4% (v/v) formaldehyde.

4. Diatoms treatment

Diatoms can be identified to species level though morphological features of their frustules but these will hardly be seen without a previous treatment consisting of making all the organic matter disappear, leaving their siliceous frustule empty. There are different techniques for diatom cleaning but the hydrogen peroxide method (30% H_2O_2 solution) is the most widely used.

Firstly, if formaldehyde has been added, it should be removed. The sample has to be homogenized by shaking and a tube for centrifugation has to be filled with it. Centrifuge the sample at 2,500 rpm for 15 minutes, remove the supernatant, refill the tube with distilled water and homogenize it again. This process should be repeated three times but after the first one the supernatant must be revised to prevent any loss of diatoms.

Once the supernatant is removed for the third time, add a few drops of hydrogen peroxide to the pellet. Be careful because a lot of foam may appear! Some more peroxide has to be added carefully until half of the tube is filled with it. Then, the tube can be either left covered for several days until bubbles stop flowing or heated in a sand bath or a hotplate at about 90 °C during 1 to 4 hours. In both cases, the brownish suspension has to become whitish.

After that, add a few drops of HCl and repeat the centrifuge process previously described to remove all the hydrogen peroxide.

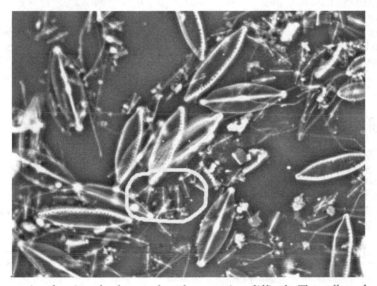

Fig. 3. An excessive density of valves makes the counting difficult. The yellow shape surrounds seven small valves of *Fistulifera saprophila* (Lange-Bertalot & Bonik) Lange-Bertalot. This species is almost impossible to detect without phase contrast. Notice that these diatoms are abundant in the sample and are widespread all over the image.

When diatoms are cleaned and suspended in water, dispense a few drops of this water on a cover slip and let it dry. Once dried, examine the cover slip under the microscope to ensure that diatoms can be easy to count. A good number of valves are eight per field but it depends on their size and shape. If there are too many, the sample must be diluted (Fig. 3); if there are only a few diatoms, add more drops and let the cover slip dry again. If there are not enough diatoms on the cover slip but there is too much inorganic matter on it, the researcher has to make a decision whether or not to add more sample to the cover slip: the addition of more sample would reduce the time of analysis but diatoms would be covered by more inorganic matter, which makes the identification of species harder.

After getting an appropriated number of diatoms on the cover slip, dispense a small drop of high refractive medium (such as Naphrax™) on a slide and place the cover slip on it, making sure that diatoms are in contact with the drop. Heat the preparation for about one minute; some bubbles will appear under the cover slip when the drop of high refractive medium boils but they will disappear as the slide cools again. Once cold, the cover slip should be closely adhered to the slide; if not, the preparation has to be heated again.

5. Diatom analysis

The next step is to identify diatoms from the sample. Some investigators obtained good results identifying to the genus level (Wu & Kow, 2002) either when samples had only a few species of each genus (Growns, 1999) or when a quick assessment of ecological quality is required (Rimet & Bouchez, 2011). However, we recommend reaching to species level when the goal of the study is one of the following: To obtain an accurate bio-assessment, to apply diatom-based water quality indexes, or to define the ecological status of a body of water. Reaching the species level the first time that samples are analysed is time-consuming but provides significantly more information than a genus-level analysis. It also permits data aggregation in higher taxonomical groups if necessary. If other than species level is reached, the information obtained is likely to be too poor, thus forcing researchers to analyse the samples again to identify and count species. In the end, this task would be much more time-consuming.

There are genera with a large number of species and these species are water quality indicators that should be taken into consideration when analysing the samples.

For example, the big genus *Nitzschia* Hassall comprises species such as *Nitzschia fonticola* Grunow (Fig. 4), which is typically present in clean water, as well as *Nitzschia capitellata* Hustedt (one of the largest species from Fig. 3), frequently found in polluted water. The observation of *Nitzschia* in one sample does not provide information neither about water quality nor the features of the ecosystems, since *Nitzschia* can be easily found in almost any water body in the world.

Therefore, it is necessary to reach to species level when calculating diatom indexes. There are some other indexes that even require variety levels. There are many previous studies that can be referenced when determining diatoms. The most widely followed are Krammer and Lange-Bertalot (1986, 1988, 1991a, b). It is useful to have more guides and bibliography when identifying diatoms but we consider these ones to be sufficient, at least as an initial approach to diatom identification.

The scientific names from these guides are not up to date but the old names of the diatoms are listed with the founding author's name and are updated in some websites such as www.algaebase.org. In relation to synonymy, there are differences in taxonomical criteria among authors; therefore, it is important referring the authors following the scientific names of diatoms. From a taxonomical point of view, the simple determination of the genus sometimes can be insufficient to ensure its own genus identity, due to the evolution of the taxonomy. For example, *Nitzschia* Hassall is split into different genera (*Nitzschia sensu stricto*, *Grunowia* Rabenhorst, *Tryblionella* W. Smith, *Psammodictyon* Mann, etc.). Knowing that a *Nitzschia* has been found is not enough to determine if that diatom is actually *Nitzschia sensu stricto* or a member of any other genus coming from *Nitzschia*, such as *Grunowia* (Fig. 5).

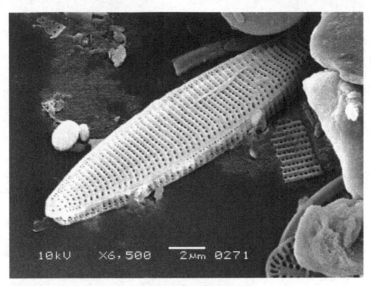

Fig. 4. External view of *Nitzschia fonticola*'s valve Grunow.

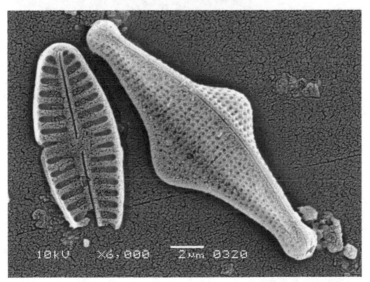

Fig. 5. The right valve is an external view of *Nitzschia sinuata* var. *tabellaria* (Grunow) Grunow, whose updated name is *Grunowia tabellaria* (Grunow) Rabenhorst.

Moreover, if species level is reached, the authors following the diatom names should be observed to update the identifications. For example, if one researcher indentifies *Nitzschia constricta*, without providing any reference regarding the authors, it would be impossible to know whether he or she has actually found *Nitzschia constricta* (Gregory) Grunow (=*Psammodictyon constrictum* (Gregory) Mann) or *Nitzschia constricta* (Kützing) Ralfs (=*Tryblionella apiculata* Gregory). A diatom analysis that determines the presence of *Nitzschia constricta* (Kützing) Ralfs gives enough evidence to indicate that we are actually referring to *Tryblionella apiculata* Gregory, in spite of not having used updated taxonomical criteria.

The diatoms must be identified at a high magnification (100X) while using immersion oil. Non-experienced researchers tend to maximize the contrast of the image observed through the microscope to detect the limits of the valves clearly but some structures, like striae, can sometimes go unnoticed because of this excessive contrast. In these cases, it would be better to increase the lighting over the sample and to reduce the contrast in order to easily observe the diatoms in detail.

It is highly recommended to combine phase contrast with bright field techniques in order to see fine structures (striae, processes, etc.). When a microscope is not equipped with a phase contrast, put an opaque object (a pencil would be enough) between the lamp and the sample, as shown in the figure 6. Light will be deflected and it would work similarly to a phase contrast. The quality of the image can be improved if the object is slightly moved, changing its orientation in relation to the observed valve. The results of this technique are not as good as they would be when using phase contrast but it will provide similar images, plus it is cheaper than purchasing a phase contrast.

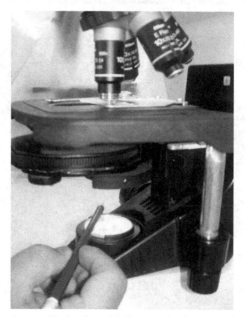

Fig. 6. An opaque object – a pencil– is placed between the lamp and the sample to simulate the phase contrast technique.

Figure 7 shows four photographs of the same valve with different types of illumination. Notice that the shape of the valve is perfectly visible in the first image, but the striae are not.

At least 400 valves have to be counted in each sample and the relative proportions of the species have to be calculated. This number provides a good estimate of the composition of the diatom flora and this is a requirement for the application of the biotic indexes.

5.1 Indexes calculation

Once the analysis is finished, the next step is to calculate biotic indexes in order to determine the water quality. Most of the widely used indexes come from the Zelinka and Marvan equation (Zelinka and Marvan, 1961), which is:

$$Index = \Sigma \ (A_i \ x \ v_i \ x \ j_i) \ / \ A_i v_i \qquad (1)$$

Where A is the relative abundance of species i, v is the indicator value of the species and j its sensitivity.

The indexes differ in both diatom species and the values v and j assigned for each index.

Existing indexes must be tested when applied to a basin different from the ones (Prygiel et al., 1999). This testing is usually done by comparing the values given by the indexes with the physicochemical data from the same sites. The Spearman correlation between an index and chemical variables (phosphate, DIN, COD) is enough to determine whether that index can be applied to the basin or not.

There are many studies regarding this issue and it has been proved that these indexes are applicable and work in different parts of the world (Torrisi & Dell'Uomo, 2006; Atazadeh et al., 2007; Taylor et al, 2007). However, most researches tend to state that it is important to test the indexes before using them to verify their accuracy (Potapova & Charles, 2007). As a matter of fact, new indexes have been developed for different basins after having found a weak correlation between chemical data and diatom indexes.

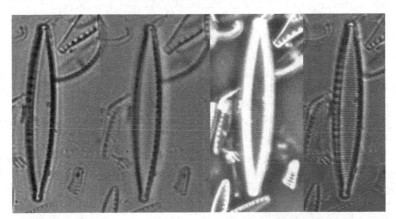

Fig. 7. Four photographs of the same valve of *Nitzschia fonticola* Grunow under different types of illumination. From left to right, bright field with a high contrast, bright field with low contrast, phase contrast, and simulation of phase contrast with an opaque object.

Index calculation is a quick and easy task if the software Omnidia, which is widely used among researchers, is available. It is an expensive computer program but it is extremely useful and worthy. It performs fast index calculations and transforms them into a scale that ranges values from 0 to 20, which facilitates comparisons. There is also a free Excel file, available at http://omnidia.free.fr/update/IBD_NEW.XLS, that permits the calculation of the IBD index. However, we recommend the acquisition of the Omnidia software.

Can diatom-based indexes be used carelessly if they correlate with the chemical data? It is not possible to do that, even though its applicability has been demonstrated all over the world. There are some situations in which indexes can present spurious values that have to be analyzed in order to avoid false conclusions. The rest of this chapter will show some examples where diatom-based indexes failed to work.

6. High index values in bad water quality and the relationship with the ecological status: The case of the Guadiamar River

This case shows a real situation in which some diatom indexes, applied in a river with poor water quality and bad ecological status, return the high values that can be found in a pristine river. When facing a similar situation, researchers must consider other ecological elements in order to determine what is actually happening in the ecosystem.

The Guadiamar River is located in the South of the Iberian Peninsula. The river has been suffering human impact for decades by receiving non-treated wastewaters, which have drastically reduced the quality of its waters. The diatom-based indexes showed low values indicating bad water quality due to the extended presence of organic matter and eutrophy in the water, as they were expected to do.

However, the Guadiamar River has also received acidic waste with a high content of heavy metals — mainly Fe, Zn, Cd, and Cu — from a nearby mine. This impact is located upstream the organic inputs. The pollution from this mine reaches the Guadiamar River through one of its tributaries, the Agrio River. This chemical pollution became more serious after a catastrofic accident in 1998, when a dam spilt tons of contaminated mud and water with high levels of heavy metals into the river.

Fortunately, the situation has improved since the accident. After hard and expensive work, today the river is a green corridor, a path communicating two different protected natural areas. Wastewater was treated in sewage treatment plants and most heavy metals were removed. Nevertheless, the water still has lower quality than desired because: 1. the eutrophy level is still too high and 2. part of the heavy metals remained within the sediment, which risk of mobilization after a big flood.

In relation to diatoms, it is interesting to analyze the composition of the flora on the areas not affected by the organic matter and eutrophy but only by the mine. We will briefly present the characteristic appearance of the river after the spilling of the mine and the subsequent cleaning, between the years 2001 to 2006. After the spill, the pH values of the water decreased and the amount of dissolved heavy metals increased drastically. This situation worsened during summer because in this season the water levels diminish, increasing the concentration of pollutants and changing the water pH to below 5.

Benthic algae were sampled in the Agrio and Guadiamar Rivers following the previously explained procedure. In the Guadiamar River, sampling was carried out upstream, in relation to the mine, in the confluence of the two rivers, and downstream.

The samples obtained upstream from the mine spilling showed small cyanobacteria, different green algae and many diatoms from different genera (*Melosira* Agardh, *Ulnaria* (Kützing) Compère, *Achnanthidium* Kützing, *Planothidium* Round et Bukhtiyarova, *Cocconeis*

Ehrenberg, *Cymbella* Agardh, *Encyonema* Kützing, *Navicula* Bory, *Gomphonema* Ehrenberg, *Nitzschia* Hassall, *Tryblionella* W. Smith and others). After the confluence of the Agrio River and the Guadiamar, the algal flora was completely different. Researchers observed a strong development of filamentous green algae *Klebsormidium* Silva, Mattox & Blackwell, *Ulothrix* Kützing and *Mougeotia* Agardh in the acidic waters along with large quantities of diatoms belonging to only a few species. The most characteristic species from the flora in this part of the river were *Brachysira vitrea* (Grunow) Ross and different taxa belonging to the complex of *Achnanthidium minutissimum* (Kützing) Czarnecki

Some diatom indexes where calculated in all these sites using the software Omnidia.

Table 1 shows the results of one sampling performed in April 2006, where the indexes suggest that both Guadiamar and Agrio rivers had good water quality. In terms of organic matter or eutrophy, the information given by the indexes was accurate. But we cannot consider the general quality of the water to be good, since it was highly polluted by acidity and metals.

Some features of the flora were more indicative of the actual state of the river. The observed green masses of *Klebsormidium* or *Ulothrix* are typical of acidic water with high levels of metals and are often found in these types of ecosystems (Stevens et al., 2001; Niyogi et al., 2002; Martín et al. 2004). Moreover, the richness of diatoms (see number of species in Table 1) was much lower in the Agrio River and downstream of the confluence of both rivers than in the upper part, upstream of the mine.

The genus *Achnanthidium* Kützing, very abundant in these rivers, is also noteworthy. This genus comprises small pennate diatoms. The most relevant ecological feature of these diatoms is an extraordinary ability to colonize, as a pioneering species, any system submitted to some kind of perturbation that leads to the desaparition of the flora of a site. This perturbation can be natural, like a flood after a heavy rain. Likewise, this genus can dominate in some samples just because the biofilm is not mature (Ács et al., 2004).

Moreover, these diatoms can live under a wide range of environmental conditions and resist different types of pollutants. On one hand, *Achnanthidium saprophilum* (Kobayasi & Mayama) Round & Bukhtiyarova was first described in waters polluted with organic matter (Kobayasi & Mayama, 1982), whereas *Achnanthidium minutissimum* (Kützing) Czarnecki (Fig. 8) and *A. biassolettianum* (Grunow) Round & Bukhtiyarova are considered indicative of clean water by the diatom indexes. The exact differentiation among these species and some others is often difficult (Potapova & Hamilton, 2007; Fig. 9) and their utilization in the indexes can sometimes be misleading (Martín et al., 2010).

On the other hand, different authors have found these diatoms growing in waters polluted by metals, as in the Guadiamar River (Sabater, 2000; Seguin et al., 2001; Szabó et al., 2005; Luís et al., 2011). Actually, this genus is frequently seen in this basin, independent of the presence of the mine, since it was abundant upstream. However, at the same time, this genus is the one that better resists the inputs from the mine. In fact, not only did *Achnanthidium* not disappear from the affected sites – as did *Amphora pediculus* (Kützing) Grunow, for example –, but also grew more than any other diatom. This developing can be linked to the fact that *Achnanthidium* may not be favoured by heavy metals and acidity but less disfavoured than other diatoms. Thus, *Achnanthidium* can grow without almost any competitors and dominate the biofilm. This could be the reason why some authors even consider *Achnanthidium* to be an indicator of heavy metal pollution (Nakanishi et al., 2004), although *Achnanthidium* is not exclusive to these environments.

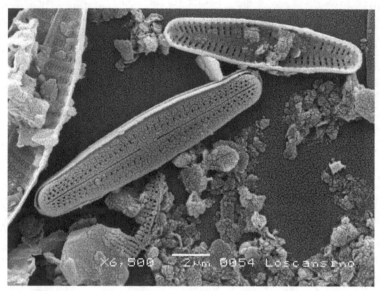

Fig. 8. Internal and external view of two valves of *Achnanthidium minutissimum* (Kützing) Czarnecki.

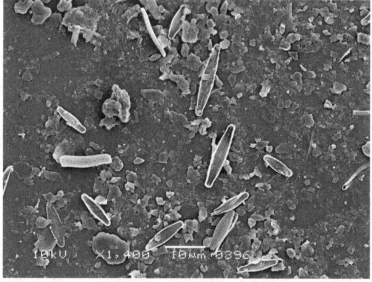

Fig. 9. A group of valves of *Achnanthidium*. Despite the different sizes, they might belong to the same species.

SPECIES	Upstream of the mine	Agrio River	Confluence of both rivers	Downstream of the confluence
Achnanthidium affine (Grunow) Czarnecki	0,0	0,0	4,8	0,0
Achnanthidium jackii Rabenhorst	3,5	0,0	21,6	1,1
Achnanthidium minutissimum (Kützing) Czarneki	30,4	52,7	0,3	0,0
Achnanthidium saprophilum (Kobayasi et Mayama) Round & Bukhtiyarova	0,0	34,4	9,3	73,9
Complex of *Achnanthidium minutissimum*	**33,9**	**87,0**	**35,9**	**75,0**
Amphora pediculus (Kützing) Grunow	22,3	0,0	0,9	0,0
Amphora veneta Kützing	0,0	0,0	2,4	0,0
Brachysira vitrea (Grunow) Ross	0,0	10,9	30,2	0,0
Cocconeis pediculus Ehrenberg	2,0	0,0	0,3	0,0
Cocconeis placentula var. euglypta (Ehrenberg) Grunow	6,3	0,0	0,0	0,0
Cyclotella meneghiniana Kützing	2,2	0,0	0,3	0,2
Cymbella amphicephala Naegeli	0,0	0,2	9,3	0,2
Encyonopsis microcephala Krammer	4,1	0,0	0,0	0,0
Eolimna minima (Grunow) Lange-Bertalot	0,0	0,0	0,3	19,8
Gomphonema angustum Agardh	0,0	0,0	2,1	0,0
Gomphonema olivaceum (Hornemann) Brèbisson	6,5	0,0	0,3	0,0
Nitzschia palea (Küt.) W. Smith	0,0	0,0	2,1	0,0
Staurosirella pinnata (Ehrenberg) Williams & Round	2,6	0,0	0,0	0,0
Ulnaria acus (Kützing) Aboal	0,0	0,0	2,1	0,0
Total number of species	**42**	**10**	**45**	**16**
IPS	**16.4**	**16.6**	**15.9**	**11.0**
IBD	**13.2**	**13.6**	**10.6**	**6.7**
EPI-D	**14.9**	**17.5**	**12.8**	**12.3**

Table 1. Species found in the river Guadiamar and Agrio during April 2006 and their percentages. Only species that contribute to the 80% of the total of diatoms in any site are included in the table and these contributions are shown in the grey cells. The percentages of the species belonging to the complex of *Achnanthidium minutissimum* are added and shown, as well as the total number of species and the values of IPS, IBD and EPI-D indexes. The colors indicate the category of water quality determined by the indexes, according to this key: red, very poor; orange, poor; yellow, moderate; green, good; blue, excellent.

A large abundance of deformed valves also indicates that the river was affected by heavy metals (Fig. 10). Some authors have also found these features in diatoms living under high levels of metals (Morín et al., 2008; Tapia, 2008).

This example shows how indexes can sometimes led us to inaccurate conclusions in situations in which pollutants are not nutrients or biodegradable organic matter. In the case of pollution by heavy metals, other features of the flora (growing of green macroalgae resistant to this type of pollution, low richness and diversity, abundance of tolerant diatoms, high percentage of deformed valves, etc.) are more reliable than biotic indexes to detect the impacts of the pollutants.

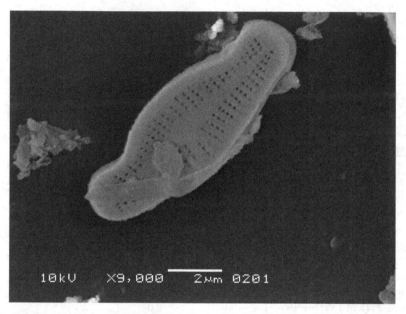

Fig. 10. A deformed valve of *Achnanthidium* from the Guadiamar River, downstream from the mine.

7. Low indexes values but good ecological status: The case of El Picacho pond

This second case study shows the opposite situation with respect to the previously analysed case study. The diatom-based indexes were high in the Guadiamar and Agrio Rivers despite the very poor water quality. In this second case, we find an ecosystem with a very good ecological status that can be underestimated by diatom indexes.

In general, a good ecological status of a water body requires good water quality. Nevertheless, some water bodies pass through different periods during its annual succession. One period can include, without man's influence, high levels of eutrophy, thus giving bad values of water quality by diatom indexes. But if it concerns the nature of the ecosystem, we cannot consider that it has a poor ecological status.

We illustrate this point with the following example. There is a small pond in the mountainous areas in the South of Spain called El Picacho (Fig. 11). It is located in a protected area (Los Alcornocales Natural Park) and has almost no human impact. Its flora and fauna are basically as they are supposed to be under no human pressures. Every spring, its crystalline waters are full of submerged plants, mainly *Myriophyllum alterniflorum* de Candolle and *Ranunculus peltatus* Schrank, the latter covering the whole pond with little beautiful white flowers. The growing of the submerged vegetation is so large that, at the end of the spring, there are not open waters at all.

The pond starts to dry when summer comes until there is no water left in the middle of the season. During the drying, the vegetation starts to die. Decaying plants produce an elevation of the organic matter and eutrophy, and the algal species that develop are indicative of this situation.

Fig. 11. El Picacho pond in August 2011.

We sampled the pond twice during summer 2011, in June and August. The most abundant submerged species was *Myryiophyllum alterniflorum* De Candolle, so it was the chosen substrate for sampling diatoms. The sampling procedure and the work carried out in the laboratory were the same as previously described in this chapter.

Table 2 shows the species found in each sample, the values of the biotic indexes IPS, IBD and EPI-D, and the water quality class determined by the indexes. As in the anterior case, the indexes were calculated using the Omnidia software.

The pond does not suffer any human impact. In this sense, it can be considered as a reference site for other small Mediterranean ponds with a similar annual succession. If the pond can be considered a reference site because of having no impacts, one could expect to have high values of the indexes (green or blue), but the indexes do not actually have a good or excellent rating. In June, when the drying has just started, the values are good (green) for EPI-D and moderate (yellow) for IPS and IBD, although they are in the limit of the poorer category (yellow). However, in August, when there is only a little water left in the pond, the values of the indexes are worse.

According to this, El Picacho pond does not achieve the European Water Framework Directive requirements, at least with respect to this indicator. But the truth is that the pond has a high ecological status because of its natural, mostly untouched condition.

This does not mean that the indexes do not work. In fact, they do work, since the water during this season is actually eutrophic. The values of the indexes, if desired to determine ecological status, have to be analysed in relation to the ecosystem characteristics and nature so as not to reach a misleading conclusion.

SPECIES	% JUNE	% AUGUST
Gomphonema gracile Ehrenberg	20.3	1.1
Gomphonema parvulum (Kützing) Kützing	29.8	9.4
Nitzschia gracilis Hantzsch	17.9	71.2
Nitzschia palea (Kutzing) W.Smith	2.1	6.8
Ulnaria acus (Kützing) Aboal	8.8	3.2
IPS	12.3	9.9
IBD	10.5	9.5
EPI-D	13.2	11.9

Table 2. Species found in El Picacho during summer 2011 and the respective percentages of each. The species that contribute less to the 5% of the total in both June and August have been deleted in order to reduce the table size. The values of IPS, IBD and EPI-D have been calculated using the Omnidia software. Colours indicate the category of water quality determined by the indexes, according to this key: Red equals very poor quality, orange-poor, yellow-moderate, green-good, and blue-excellent.

8. Conclusion

The diatom-based indexes are widely used and have proved to work in many areas of the world. However, they should be tested before being applied in a basin that was never previously studied. They are mainly used to detect organic pollution and eutrophy.

The sampling procedures to ensure a good calculation, treatment and analysis of the samples are normalised.

However, these indexes are only a working tool, meaning that the information that they provide should be compared to other sources of information from the same sites, such as chemical analysis, diversity of diatoms, and other elements of the flora, among others. Otherwise, indexes could be misinterpreted when estimating water quality and the ecological status of an ecosystem.

9. Acknowledgements

We thank Mr. Andrés Martín and and Ms. Kelly C. Korpiel for their valuable comments on the manuscript as well as their kind review of our English.

10. References

Ács, E.; Szabó, K.; Tóth, B. & Kiss, T. (2004). Investigation of benthic algal communities in connection with reference conditions of the Water framework Directives. *Acta Botanica* Hungarica, Vol. 46, N° 3-4, pp. 255-277.

Atazadeh, I.; Sharifi, M. & Kelly, M.G. (2007) Evaluation of the Trophic Diatom Index for assessing water quality in River Gharasou, western Iran. *Hydrobiologia*, 589: 165-173.

Blanco, S., Ector, L & Becares, E. (2004). Epiphytic diatoms as water quality indicators in Spanish shallow lakes. *Vie et Milieu*, Vol. 54, pp. 71-79.

Cejudo-Figueiras, C.; Álvarez-Blanco, I.; Bécares, E. & Blanco, S. (2010). Epiphytic diatoms and water quality in shallow lakes: the neutral substrate hypothesis revisited. *Marine and Freshwater Research*, Vol. 61, pp. 1457-1467.

European Standard EN 13946. (2003). Water quality – Guidance Standard for the routine sampling and pretreatment of benthic diatoms from rivers. European Committee for Standarization, Brussels

Growns, I. (1999). Is genus or species identification of periphytic diatoms required to determine the impacts of river regulation? *Journal of Applied Phycology*, Vol. 11, pp. 273-283.

Hering, D.; Johnson, R.K.; Kramm, S.; Schmutz, S. ; Szoszkiewicz, K. & Verdonschot, P.F.M. (2006). Assessment of European streams withn diatoms, macrophytes, macroinvertebrates and fish: a comparative metric-based analysis of organism response to stress. *Freshwater Biology*, Vol. 51, pp. 1757-1785.

Kelly, M.G.; Cazaubon, A.; Coring, E.; Dell'Uomo, A.; Ector, L.; Goldsmith, B.; Guasch, H.; Hürlimann, J.; Jarlman, A.; Kawecka, B.; Kwandrans, J.; Laugaste, R.; Lindstrom, E.A.; Leitao, M.; Marvan, P.; Padisák, E.; Pipp, E.; Prygiel, J.; Rott, E.; Sabater, S.; van Dam, H. & Vizinet, J. (1998). Recommendations for the routine sampling of diatoms for water quality assessments in Europe. *Journal of Applied Phycology*, Vol. 10, pp. 215-224.

Kobayasi, H. & Mayama, S. (1982) Most pollution-tolerant diatoms of severly polluted rivers in the vecinity of Tokyo. *Japanese Journal of Phycology*, Vol. 30, pp. 188-196.

Krammer, K. & Lange-Bertalot, H. (1986). *Süsswasserflora von Mitteleuropa. Band 2/1. Bacillariophyceae. Naviculaceae.* Stuttgart: Gustav Fischer.

Krammer, K. & Lange-Bertalot, H. (1988). *Süsswasserflora von Mitteleuropa. Band 2/2. Bacillariophyceae. Epithemiaceae, Bacillariaceae, Surirellaceae.* Stuttgart: Gustav Fischer.

Krammer, K. & Lange-Bertalot, H. (1991a). *Süsswasserflora von Mitteleuropa. Band 2/1. Bacillariophyceae. Centrales, Fragilariaceae, Eunotiaceae.* Stuttgart: Gustav Fischer.

Krammer, K. & Lange-Bertalot, H. (1991b). *Süsswasserflora von Mitteleuropa. Band 2/1. Bacillariophyceae. Achnanthaceae. Kritische Ergänzungen zu Navicula (Lineolatae) und Gomphonema. Gesamtliteraturverzeichnis für Teil 1-4.* Stuttgart: Gustav Fischer.

Luís, A.T.; Teixeira, P.; Almeida, S.F., Matos, J.X. & da Silva, E.F. (2011). Environmental impact of mining activities in the Lousal area (Portugal): chemical and diatom characterization of metal-contamined stream sediment and surface water of Corona stream. *Science of the Total Environment,* Vol. 409, N° 20, pp. 4312-25.

Martín, G.; Alcalá, E.; Solá, C.; Plazuelo, A., Burgos, M.D.; Reyes, E. & Toja, J. (2004). Efecto de la contaminación minera sobre el perifiton del río Guadiamar. *Limnética,* Vol. 23, pp. 315-330.

Martín, G.; Toja, J.; Sala, S.E.; Fernández, R.; Reyes, I. & Casco, M.A. (2010). Application of diatom biotic indexes in the Guadalquivir River Basin, a Mediterranean basin. Which one is the most appropriated? *Environmental Monitoring and Assessment,* Vol. 170, pp. 519-534.

Morín, S.; Duong, T.T.; Dabrin, A.; Coynel, A.; Herlory, O.; Baudrimont, M.; Delmas, F.; Durrieu, G.; Schäfer, J.; Winterton, P.; Blanc, G. & Coste, M. (2008). Long-term survey of heavy-metal pollution, biofilm contamination and diatom community structure in the Riou Mort watershed, South-West France. *Environmental Pollution,* Vol. 151, N° 3, pp. 532-542.

Nakanishi, Y.; Sumita, M.; Yumita, K.; Yamada, T.; Honjo, T. (2004). Heavy-metal pollution and its state in algae in Kakehashi River and Godani River at the foot of Ogoya mine, Ishikawa Prefecture. *Analytical Sciences,* Vol. 20, N° 1, pp. 73-78.

Niyogi, K.D.; Lewis, W.M. & McKnight. (2002). Effects of stress from mine drainage diversity, biomass and function of primary producers in mountain streams. *Ecosystems,* Vol. 5, pp. 564-567.

Potapova, M. & Charles, D.F. (2005). Choice of substrate in algae-based water-quality assessment. *Journal of the North American Benthological Society,* Vol. 24, N° 2, pp. 415-427.

Potapova, M. & Charles, D.F. (2007). Diatom metrics for monitoring eutrophication in rivers of the United States. *Ecological Indicators,* Vol. 7, pp. 48-70.

Potapova, M. & Hamilton, P.B. (2007). Morphological and ecological variation within the *Achnanthidium minutissimum (Bacillariophyceae)* species complex. *Journal of Phycology,* Vol. 43, pp. 561-565.

Prygiel, J.; Coste, M. & Bukowska, J. (1999). Review of the major diatom-based techniques for the quality assessment of rivers – state of the art in Europe, In: *Use of algae for monitoring rivers III,* J. Prygiel, B.A. Whitton & J. Bukowska, (Eds.), 224-238.

Rimet, F. & Bouchez, A. (2011). Biomonitoring river diatoms: implications of taxonomic resolution. *Ecological indicators*, Vol. 15, pp. 92-99.

Round, F.E. (1991). Use of diatoms for monitoring rivers, In: *Use of algae for monitoring rivers*, B.A. , E. Rott & G. Friedrich, (Eds.), 25-32, Düsseldorf.

Sabater, S. (2000). Diatom communities as indicators of environmental stress in the Guadiamar River, S-W Spain, following a major mine tailing spill. *Journal of Applied Phycology*, Vol. 12, pp. 113-124.

Seguin, F.; Druart, J.C.; Le Cohu, R. (2001). Effects of atrazine and nicosulfuron on periphytic diatom communities in freshwater outdoor lentic mesocosm. *Annales de Limnologie*, Vol. 37, pp. 3-8.

Sekar, R.; Venugopalan, V.P.; Nandakumar, K.; Nair, K.V.K & Rao, V.N.R. (2004) Early stages of biofilm sucesión in a lentic freshwater environment. *Hydrobiologia*, Vol. 512, pp. 97-108.

Stevens, A.E.; McCarthy, B.C. & Vis, M.L. (2001). Metal content of *Klebsormidium*-dominated (Chlorophyta) algal mats from acid mine drainage waters in southeastern Ohio. *Jounal of the Torrey Botanical Society*, Vol. 128, N° 3, pp. 226-233.

Szabó, K.; Kiss, K.T.; Taba, G. & Ács, E. (2005). Epiphytic diatoms of the Tisza River, Kisköre Reservoir and some oxbows of the Tisza River after the cyanide and heavy metal pollution in 2000. *Acta Botanica Croatica*, Vol. 64, N° 1, pp. 1-46.

Tapia, P.M. (2008). Diatoms as bioindicators of pollution in the Mantaro River, Central Andes, Peru. *International Journal of Environment and Health*, Vol. 2, N° 1, pp. 82-91.

Taylor, J.C.; Prygiel, J.; Vosloo, A.; de la Rey, P.A. & van Rensburg, L. (2007). Can diatom based pollution indexes be used for biomonitoring in South Africa? A case of study of The Crocodile West and Marico water management area. *Hydrobiologia*, Vol. 592, pp. 455-464.

Torrisi, M. & Dell'Uomo, A. (2006). Biological monitoring of some Apennine rivers (Central Italy) using the Diatom-based Eutrophication/Pollution Index (EPI-D) compared to other European diatom indexes. *Diatom Research*, Vol. 592, pp. 159-174.

Torrisi, M.; Scuri, E.; Dell'uomo, A. & Cocchioni, M. (2010). Comparative monitoring by means of diatoms, macroinvertebrates and chemical parameters of an Apennine watercourse of central Italy: the River Tenna. *Ecological Indicators*, Vol. 10, N° 4, pp. 910-913

Towsend, S.A & Gell, P. (2005). The role of substrate type on benthic diatom assemblages in the Daly and Roper Rivers of the Australian Wet/Dry Tropics. *Hydrobiologia*, Vol. 548, pp. 101-115.

Triest, L.; Kaur, P.; Heylen, S. & de Pauw N. (2001). Comparative monitoring of diatoms, macroinvertebrates and macrophytes in the Woluwe River (Brussels, Belgium). *Aquatic Ecology*, Vol. 35, N° 2, pp. 183-194.

Wu, J.T. & Kow, L.T. (2002). Applicability of a generic index for diatom assemblages to monitor pollution in the tropical river Tsanwun, Taiwan. *Journal of Applied Phycology*, Vol. 14, pp. 63-69.

Zelinka, M. & Marvan, p. (1961). Zur Prazisierung der biologischen Klassification des Reinheit fliessender Gewasser. *Archiv fur Hydrobiologie*, Vol. 57, pp. 389-407.

Permissions

The contributors of this book come from diverse backgrounds, making this book a truly international effort. This book will bring forth new frontiers with its revolutionizing research information and detailed analysis of the nascent developments around the world.

We would like to thank Kostas Voudouris and Dimitra Voutsa, for lending their expertise to make the book truly unique. They have played a crucial role in the development of this book. Without their invaluable contribution this book wouldn't have been possible. They have made vital efforts to compile up to date information on the varied aspects of this subject to make this book a valuable addition to the collection of many professionals and students.

This book was conceptualized with the vision of imparting up-to-date information and advanced data in this field. To ensure the same, a matchless editorial board was set up. Every individual on the board went through rigorous rounds of assessment to prove their worth. After which they invested a large part of their time researching and compiling the most relevant data for our readers. Conferences and sessions were held from time to time between the editorial board and the contributing authors to present the data in the most comprehensible form. The editorial team has worked tirelessly to provide valuable and valid information to help people across the globe.

Every chapter published in this book has been scrutinized by our experts. Their significance has been extensively debated. The topics covered herein carry significant findings which will fuel the growth of the discipline. They may even be implemented as practical applications or may be referred to as a beginning point for another development. Chapters in this book were first published by InTech; hereby published with permission under the Creative Commons Attribution License or equivalent.

The editorial board has been involved in producing this book since its inception. They have spent rigorous hours researching and exploring the diverse topics which have resulted in the successful publishing of this book. They have passed on their knowledge of decades through this book. To expedite this challenging task, the publisher supported the team at every step. A small team of assistant editors was also appointed to further simplify the editing procedure and attain best results for the readers.

Our editorial team has been hand-picked from every corner of the world. Their multi-ethnicity adds dynamic inputs to the discussions which result in innovative outcomes. These outcomes are then further discussed with the researchers and contributors who give their valuable feedback and opinion regarding the same. The feedback is then collaborated with the researches and they are edited in a comprehensive manner to aid the understanding of the subject.

Apart from the editorial board, the designing team has also invested a significant amount of their time in understanding the subject and creating the most relevant covers. They scrutinized every image to scout for the most suitable representation of the subject and create an appropriate cover for the book.

The publishing team has been involved in this book since its early stages. They were actively engaged in every process, be it collecting the data, connecting with the contributors or procuring relevant information. The team has been an ardent support to the editorial, designing and production team. Their endless efforts to recruit the best for this project, has resulted in the accomplishment of this book. They are a veteran in the field of academics and their pool of knowledge is as vast as their experience in printing. Their expertise and guidance has proved useful at every step. Their uncompromising quality standards have made this book an exceptional effort. Their encouragement from time to time has been an inspiration for everyone.

The publisher and the editorial board hope that this book will prove to be a valuable piece of knowledge for researchers, students, practitioners and scholars across the globe.

List of Contributors

Neslihan Balkis and Benin Toklu-Aliçli
Istanbul University, Faculty of Science, Department of Biology, Vezneciler-Istanbul, Turkey

Muharrem Balci
Istanbul University, Institute of Science, Vezneciler-Istanbul, Turkey

Ch. Ntislidou, A. Basdeki, Ch. Papacharalampou, K. Albanakis, M. Lazaridou and K. Voudouris
Interdisciplinary Postgraduate Study Program "Ecological Water Quality and Management at a River Basin Level", Departments of Biology, Geology & Civil Engineering, Aristotle University of Thessaloniki, Thessaloniki, Greece

Cesar João Benetti, Amaia Pérez-Bilbao and Josefina Garrido
University of Vigo, Spain

Anca Farkas and Dorin Ciataraş
Someş Water Company, Cluj-Napoca, Romania

Brânduşa Bocoş
National Public Health Institute, Regional Public Health Center of Cluj, Cluj-Napoca, Romania

Radovan Kopp
Mendel University in Brno, Department of Fisheries and Hydrobiology, Czech Republic

O. De Castro
Department of Biological Sciences-University "FedericoII" of Naples, Italy

M. Inglese
Laboratory of Environmental and Food Analyses Studies "Inglese & co. ltd", Italy

G.D'Acunzi and F. Landi
Salerno District Administration Bureau, Italy

M. Guida, S. Leva and L. Copia
Department of Structural and Functional Biology, University "Federico II" of Naples, Italy

R.A. Nastro
Department of Sciences for the Environment, University Parthenope of Naples, Italy

Raquel B. Queirós
INESC TEC (formerly INESC Porto) and Faculty of Sciences, University of Porto, Portugal
BioMark Sensor Research and ISEP/IPP - School of Engineering, Polytechnic Institute of
Porto, Portugal

P.V.S. Marques
INESC TEC (formerly INESC Porto) and Faculty of Sciences, University of Porto, Portugal

M. Goreti F. Sales
BioMark Sensor Research and ISEP/IPP - School of Engineering, Polytechnic Institute of
Porto, Portugal

J.P. Noronha
REQUIMTE/CQFB and Faculty of Sciences and Technology of University Nova de Lisboa,
Portugal

Lila Ferrat, Christine Pergent-Martini, Gérard Pergent and Vanina Pasqualini
University of Corsica, Sciences for Environment, France

Sandy Wyllie-Echeverria
Friday Harbor Laboratories, University of Washington USA

G. Cates Rex and Jiping Zou
Brigham Young University, Department of Biology, USA

Michèle Romeo
University of Nice Sophia-Antipolis, ROSE, France

Catherine Fernandez
University of Aix-Marseille, IMEP, France

Suman Manna, Kaberi Chaudhuri, Kakoli Sen Sarma, Pankaj Naskar and Somenath Bhattacharyya
Institute of Environmental Studies and Wetland Management, Kolkata, India

Maitree Bhattacharyya
Department of Biochemistry, University of Calcutta, Kolkata, India

Gonzalo Martín and María de los Reyes Fernández
University of Seville, Spain

Printed in the USA
CPSIA information can be obtained
at www.ICGtesting.com
JSHW011431221024
72173JS00004B/765